On
the Human Condition

'Dominique Janicaud was one of the French philosophers most attentive to contemporary realities and their origins, without ever forfeiting conceptual rigour.'

Roger Pol-Droit, *Le Monde*

Praise for the series

'... allows a space for distinguished thinkers to write about their passions.'

The Philosophers' Magazine

'... deserve high praise.'

Boyd Tonkin, The Independent (UK)

'This is clearly an important series. I look forward to receiving future volumes.'

Frank Kermode, author of Shakespeare's Language

'... both rigorous and accessible.'

Humanist News

'... the series looks superb.'

Quentin Skinner

'... an excellent and beautiful series.'

Ben Rogers, author of A.J. Ayer: A Life

'Routledge's Thinking in Action series is the theory junkie's answer to the eminently pocketable Penguin 60s series.'

Mute Magazine (UK)

'Routledge's new series, Thinking in Action, brings philosophers to our aid ...'

The Evening Standard (UK)

'... a welcome new series by Routledge.'

Bulletin of Science, Technology and Society (Can)

DOMINIQUE JANICAUD

Translated by Eileen Brennan

With an introduction by
Simon Critchley

On
the Human Condition

Liberté • Égalité • Fraternité
RÉPUBLIQUE FRANÇAISE

Routledge
Taylor & Francis Group

LONDON AND NEW YORK

'On the Human Condition' first published 2002 in French as
'L'homme va-t-il dépasser l'humain?'
by Bayard, 3 et 5 rue Bayard, 75008 Paris, France

English translation as 'On the Human Condition' first published 2005
by Routledge
2 Park Square, Milton Park, Abingdon, Oxon OX14 4RN

Simultaneously published in the USA and Canada
by Routledge
270 Madison Ave, New York, NY 10016

Routledge is an imprint of the Taylor & Francis Group

'L'homme va-t-il dépasser l'humain?' © 2002 Bayard
'On the Human Condition' English translation © 2005 Routledge
Introduction © 2005 Simon Critchley

*This book is supported by the French Ministry for Foreign Affairs, as
part of the Burgess programme headed for the French Embassy in
London by the Institut Français du Royaume-Uni*

Typeset in Joanna MT and DIN by
RefineCatch Ltd, Bungay, Suffolk
Printed and bound in Great Britain by
TJ International Ltd, Padstow, Cornwall

British Library Cataloguing in Publication Data
A catalogue record for this book is available from the British Library

Library of Congress Cataloging-in-Publication Data
Janicaud, Dominique, 1937–
    [Homme va-t-il dépasser l'humain. English]
    On the human condition / Dominique Janicaud.
        p. cm. — (Thinking in action)
    Includes bibliographical references.
    1. Philosophical anthropology.   2. Superman (Philosophical concept)
    3. Biotechnology—Philosophy.   4. Humanism.   I. Title.   II. Series.
B2430.J283H6613 2005
128—dc22   2005001403

ISBN 0–415–32795–4 (hbk)
ISBN 0–415–32796–2 (pbk)

Contents

Introduction: The Overcoming of Overcoming
On Dominique Janicaud
Simon Critchley   vii

Preface   **1**

Is Humanism the Last Resort?   **One**   **5**

The Danger of Monsters   **Two**   **19**

**From Foreseeable Risks to the
Unforeseeable**   **Three**   **35**

Between the Superhuman and the Inhuman   **Four**   **44**

Conclusion: What 'Overcoming' Means   **54**

Notes   59
Bibliography   66
Index   69

Introduction: The Overcoming of Overcoming

On Dominique Janicaud

Simon Critchley

The book by Dominique Janicaud that you have in your hands appeared a few months after his untimely death in August 2002.[1] The original French title of *On the Human Condition* took the form of a question: *L'homme va-t-il dépasser l'humain?*[2] (*Will Man Overcome the Human?*) For reasons that I hope will soon become clear, the word that provides my focus in this introduction is the verb *dépasser*, to overcome, and the related substantive *dépassement*, overcoming. In this way, I hope to bring out a recurrent feature of Janicaud's work, and what is arguably its governing logic. In my view, the overwhelming *critical* intention of Janicaud's work is to leave behind all fantasies of overcoming, whether that concerns an overcoming of metaphysics, of rationality, or humanity as such. Renouncing such fantasies, which recur with frightening regularity – in the 1980s and 1990s around the question of artificial intelligence and more recently in bio-ethical debates on genetic modification, mutation and cloning – Janicaud's sage counsel is to attain what I would like to call an 'overcoming of overcoming'. That is, to leave behind all apocalyptic discourse on the end, whether the end of man, of history or whatever, and all concomitant talk of a new beginning, of the post-human or post-history.

As Janicaud makes clear in 'Heideggeriana', a fragmentary meditation from the early 1980s which echoes the form and content of Heidegger's own collection of fragments 'Überwinding der Metaphysik' ('Overcoming Metaphysics'), the idea of an overcoming of overcoming is inherited from Heidegger.[3] In almost the final words of the important 1961 lecture, *Zeit und Sein*, Heidegger writes: 'Yet a regard for metaphysics still prevails even in the intention to overcome metaphysics. Therefore, our task is to cease all overcoming, and leave metaphysics to itself'.[4] However, as we will see, Janicaud's understanding of these words is not Heideggerian in any orthodox sense, but on the contrary opens up a new possibility for thinking about reason and rationality that refuses the opposition between metaphysics, on the one hand, and meditative thinking or what Heidegger calls *Gelassenheit*, on the other. I would like to begin by trying to clarify Janicaud's line of interrogation with respect to metaphysics and its overcoming, before going on to discuss how this decisively influences his innovative approach to rationality. In conclusion, I will try to spell out the vision of the human that might be said to follow once one has attained an overcoming of overcoming. To that extent, *On the Human Condition* is not some afterthought or appendix to Janicaud's work, but rather the extension of its logic into the question of the meaning of the human. Indeed, it might be seen as a conclusion of sorts.

*

In 1973, Janicaud published 'Dépasser la métaphysique?' ('Overcoming Metaphysics?'), a title where what should be emphasized is the sceptical question mark. After a careful identification of the different strands of Heidegger's strategies with regard to metaphysics, Janicaud adds the following revealing remark in a postscript that was written for the essay's republication in 1983: 'Formally, we can claim at once that metaphysics is overcome by Heidegger ... and that it is acknowledged to have an unsurpassable character.'[5] Or again, 'Delimited, left to itself, metaphysics can continue to exist.'[6]

What Janicaud identifies in Heidegger is what he calls on several occasions 'the aporia of overcoming' ('l'aporie du dépassement'). Now, this aporia or perplexity has a consequence that is both Heideggerian and anti-Heideggerian. On the one hand, Janicaud emphasizes the uncertainty of any project of the overcoming of metaphysics in Heidegger in order to prevent the kind of misinterpretation that one finds in Deleuze (and he is not alone: one can find similar sentiments in Rorty, Habermas and many others) when he attributes to Heidegger the idea of 'an exit outside the metaphysical field' or 'a turning beyond metaphysics'.[7] All talk of the overcoming of metaphysics in Heidegger has to be linked to the idea of a *Verwindung* of metaphysics, that is, a reappropriation of metaphysics in terms of its unthought essence, what Janicaud translates as 'rémission', a sort of re-sending or repeat transmission of what Heidegger calls the original sending of being. However, on the other hand, this aporia is anti-Heideggerian insofar as Janicaud argues that it is simply false to claim that the previous, i.e. pre-Heideggerian, history of metaphysics, and he is thinking in particular of Hegel, is incapable of thinking what Heidegger called 'the truth of being'. Therefore, if it is false to claim that Heidegger believed that we could leave metaphysics behind, it is also false of Heidegger to claim that the previous history of metaphysics was unable to think the truth of being as such. Heidegger's conception of the history of metaphysics suffers from what we might call a certain 'unilateralism'. In an autobiographical text published in English in 1997, Janicaud wrote extremely candidly of his sharp disagreement with Heidegger:

I could accept neither the schema of history nor that of Being, nor the secret, destinal correspondence of the originary and the *Ereignis*. And I do not think that meditative thought can preserve a resource against technicist nihilism if it refuses all specific understanding of new realities, which always resound with ambiguity.[8]

One of the most impressive features of Janicaud's work was its detailed engagement with those new realities and he had an impressive knowledge of both the history and philosophy of science and much contemporary scientific research.

In 'Heideggeriana', Janicaud writes: 'It is therefore false to claim that metaphysics does not think ontological difference, just as it is false to understand the Heideggerian overcoming of metaphysics as a "going beyond" (*"outrepassement"*).'[9]

The philosophical consequence of the aporia of overcoming is simple, but far-reaching: it leads Janicaud to question Heidegger's separation between, on the one hand, metaphysical rationality and, on the other, the meditative thinking of being which Heidegger saw as the unthought ground of reason. That is, if metaphysics in the period of what Heidegger called its completion (*Vollendung*) continues to exist, then the task of thought is not a meditation on the truth of being, but rather a *philosophical* thinking of reason and rationality that would avoid this Heideggerian separation, a separation which risks congealing into a cleavage. In short, if we can say with Heidegger and against Carnap, Rorty and Habermas that metaphysics is not decisively overcome, then Janicaud invites us to say *against* Heidegger that rationality is not entirely containable within a reductive metaphysics whose alternative is a pre-rational experience of *Gelassenheit* or poetic meditation. On the contrary, despite the hyper-rationality of what Heidegger calls the attitude of enframing (*Gestell*) that defines the age of technology, rationality holds opens a whole domain of possibility, potential or *puissance* whose analysis is the task of philosophical *intelligence*. It is interesting to enumerate the various occasions and contexts in which Janicaud employs the word 'intelligence'. For example, in *Powers of the Rational* (*La puissance du rationnel*), he speaks of 'the intelligence of the enigma', and I will come back to the question of enigma;[10] in *Chronos*, he speaks of 'the intelligence of the temporal *partage*', and I will also come back to the meaning of the word *partage*; in *L'homme va-t-il dépasser l'humain?*, he speaks of 'the intelligence of our mortal and fragile *partage*'; and in a long posthumously published essay, 'Vers l'intelligence du partage', Janicaud speaks of 'the fleeting fragility of intelligence'.[11] It is a favourite word in his lexicon, as indeed is the word 'philosophy'. With an increasing firmness, which perhaps testifies to his ongoing debt to Hegel, Janicaud sought to defend the notion of philosophy and philosophical intelligence against the retreat of Heidegger and Heideggerians into meditative thinking, a tendency that finds its clearest expression in the 1991 collection of essays, *À nouveau la philosophie*.[12] In 'Heideggeriana', Janicaud writes: 'Breaking through the hardening dichotomy between metaphysics and the thinking of being, I would suggest that there subsists a *possibility for the rational* (un *possible rationnel*) that apportions itself in fields of intelligibility more open than operative or instrumental rationality'.[13] This project of *un possible rationnel* finds its decisive expression in what is undoubtedly Janicaud's major philosophical work, *Powers of the Rational* from 1985. Let me now turn to this book.

*

Powers of the Rational begins from the Weberian premise that we are in the grip of an aggressive and aggressively globalizing rationalization whose principal means of expansion is technologized science or what Janicaud, after Jacques Ellul, calls 'techno-science'. Such rationalization is linked both to the scientific project of the mastery of nature that has defined modernity since Descartes and Bacon, but also to the military, industrial and informational operational deployment of science through the cultivation of research and development (R&D). Janicaud's hypothesis is that today the rationality of techno-science has become what he calls '*une surrationalité*', or hyper-rationality. This is what Heidegger calls the *Gestell* or power of enframing, and what Janicaud sees as the intensification of the process of rational power whose goal is the total *actualization* or *effectuation* (*Wirklichkeit*) of the powers of the possible. As such, contemporary techno-science is characterized by a sheer wilfulness, a desire for total actualization; what Heidegger would see as 'the will-to-will', where the hyper-rationality of techno-science risks reversing itself, becoming irrational. In other words, to follow Adorno and Horkheimer, there is a dialectical inversion of the process of rational enlightenment, an irrationality linked for us to the names of Auschwitz and Hiroshima, but which equally defines the contemporary logic of corporate governance, scientific research and its technological implementation. It also defines the terrifyingly rational irrationality of our current context, which is – should anyone forget – a situation of war.

What, then, is one to do faced with the all-pervasive irrationality of rationality? Well, one option would be to follow Heidegger and argue for some sort of separation between rationality and the thinking of being or *Gestell* and *Ereignis*, but Janicaud has already excluded this option for the reasons given above. Janicaud's conviction, a conviction that I would see as Pascalian (and I will be coming back more than once to Pascal), is that we cannot take leave of rationality simply because its limits have been shown. As Pascal would put it, there are two excesses: to exclude reason and to admit nothing but reason. The task of thinking consists in trying to render intelligible the massive and inevitable presence of rationality in order to mark a limit to the irrationality of rationalization. If the latter is defined by the attempted actualization of all the powers of possibility, then the task of philosophical intelligence is to produce an account of rationality that testifies to a certain *puissance* or potency of the rational. This explains the deliberate ambiguity in the French title of *La puissance du rationnel*, where it is a question of a certain potency of rationality, *un rationnel puissant*, which is not that of the order of *Puissance* or Power with a capital 'P' that attempts to actualize the possible.

For Janicaud, scientific rationality is characterized by 'potentialization'. This word has a double meaning, being at once the enabling of the possible and the withholding of the complete effectuation of the possible in action: potentialization has to remain potent. Thus, science potentializes: it makes possible forms of human knowledge and action that were hitherto unimaginable. The Heideggerian question of whether science does or does not think is a cul-de-sac; the point is that scientific rationality makes possible new forms of human activity with higher degrees of coherence, universality and explanatory power, and to deny this is simply to fall into anti-scientific obscurantism. However, the irrationality of contemporary rationalization consists in the fact that it sacrifices this power of the possible in the name of total actualization. Thus, the irrationality of the rational consists in the privileging of the actual over the possible. The ambition of

Powers of the Rational, which brings us back once more to its title, is to leave open the space of possibility for the rational.

If scientific rationality is characterized by potentialization, then at the heart of *Powers of the Rational* is a genealogical account of the four phases of potentialization. Very roughly, these four phases might be summarized in the following terms: (i) phase one is the potentialization of technique in the power of tools which allow for technical 'know-how' in the domain of human praxis. (ii) Phase two is the potentialization enabled by the mathematical or geometrical abstraction of entities – *mathesis* – which is characterized by the work of Euclid. (iii) Phase three is the extension of the apodicticity of Greek geometry to domains that the Greeks would not have imagined possible. This is the scientific project of modernity, linked by Husserl to the name of Galileo and the mathematization of nature, and whose aim is the latter's total mastery through science. (iv) Phase four has already been partially described in terms of the reversal of rationality into irrationality and the privilege of the actual over the possible. This is something quite new in human history, where the coupling of science and technology becomes a passionate and ultimately destructive love affair. Through the power of R&D, techno-science becomes available for industrial, military and informational processes where these processes increasingly define the nature and scope of scientific research, not to mention furnishing its financial conditions of possibility.

So, can we imagine a fifth phase in this genealogy, a new potentialization or *puissance* for rationality? That is the wager of the last chapters of *Powers of the Rational*, where Janicaud sketches a more reasonable notion of reason that he calls *partage*. This word has many shades of meaning in French, denoting both sharing and division. But the sense of the word that Janicaud liked to emphasize was the idea of rationality as *notre partage*, that is, as our lot or portion. I remember suggesting to Janicaud the idea of *partage* as 'allotment', which both suggests the idea of 'our lot in life', but also the portioning out of a piece of land, a piece of ground that would be allotted to a person but still owned in common. Indeed, there was an 'allotment' movement in England from the late nineteenth century which was linked to the emergence of co-operative societies, where ordinary working people would grow their fruit and vegetables in an allotment. Thus, *partage* is our share, our lot, the small piece of time and space that we are allotted upon an earth whose ownership is held in common and held in trust. Janicaud makes a compelling distinction between *partage* and destiny: if the latter suggests a sheer necessity working itself out despite our free choice, then *partage* is the thrown and utterly contingent character of human life, what Heidegger would call 'facticity', whose understanding is the task of philosophical intelligence.[14]

Far from submitting to some finally obscurantist fantasy of an overcoming of rationality, what Janicaud was trying to think was a non-dominating, non-instrumental and dialogic experience of rationality as that which is *shared* by mortals in their everyday being-with-one-another. In many ways, Janicaud's critique of Heidegger's division between meditative thinking and technologized reason echoes Habermas's critique of Adorno's univocal notion of instrumental rationality that is opposed to aesthetic experience. However, at that point the similarities end, and unlike Habermas's rather blunt and explicitly post-metaphysical theorization of communicative action, Janicaud's conception of rationality as *partage* is presented in a much more fragile and experimental manner in a series of dialogues and philosophical experiments. For example, in *Powers of the Rational*, we are presented with a long and compelling dialogue between 'Y', a critical rationalist, 'X', a neo-Hegelian, and 'Z', who might be described as a 'Janicaudian'. Similarly, there is a wonderful dialogue, 'Heidegger à New York', between two men and two women in a loft in Manhattan. Or again we might think of the 45 meditative fragments entitled 'Chroniques' that appear as an epilogue to *Chronos*. Finally, the posthumous *Aristote aux champs-élysées* is both a series of imagined dialogues with Aristotle, Kant, Nietzsche and Heidegger and a sequence of more solitary colloquies, often tightly aphoristic and highly lyrical in style. Such texts are experiments, they are performative enactments of *partage*, which are faithful to the fleeting fragility and delicacy of philosophical intelligence.

*

How does humanity look from the perspective of *partage*? Towards the end of the dialogue between 'X', 'Y' and 'Z' in *The Powers of the Rational*, the 'Janicaudian' personage 'Z' makes the following astonishing remark:

> It is clear that everything depends upon the manner in which humanity assumes the inevitable. There could be some surprises. If rationalization is passively accepted as a necessary collective resignation in favour of more efficient organization, then we will have the worst of destinies: subjection and tyranny. If, on the contrary, rationalization is felt as a call or appeal, as a new source of creativity that our recovered energies can make use of, then *perhaps* a new clearing awaits the world, more radiant yet than its Greek model . . .[15]

Admittedly, this passage contains an important 'perhaps', it ends with a sceptical question mark, and Janicaud is not quite speaking in his own voice, but in a rather grand style. Yet what interests me here is precisely the possibility, the potential, for thinking about rationality as a call, an appeal, a new source of creativity, a human creativity that allows, in turn, for new forms of inventiveness of the human. That is, if Janicaud's overcoming of overcoming invites us to give up the fantasies of an abandonment of metaphysics or rationality, then it is also a question of giving up the fantasies of the overcoming of the human in the post-human, superman or the overman. On the contrary, it is a question of creating new possibilities or potentialities for the human: new forms of humanization. To put the point a little more polemically, as Zarathustra teaches, man is a rope fastened between animal and overman or Übermensch, a rope over an abyss. But this does not imply, as Zarathustra also teaches, that man is something to be overcome. On the contrary, what has to be overcome is the desire for overcoming itself. When we have achieved an overcoming of overcoming, then perhaps we can attend to the finally enigmatic character of the human condition, and to the utterly fragile and un-heroic nature of that condition. The human being is not something to be overcome, but undergone. We can take the piece of rope that we are and choose to hang ourselves with it, or at least try to do so and fail, as in Beckett's Waiting for Godot. However, we can also take the rope in our hands, stretch it tight between animal and overman and try to find our feet, find our balance, and find our way.

As I read it, this is the lesson of Janicaud's On the Human Condition, a book that finally owes more to Pascal than to Nietzsche. This book is a Zeitdiagnose, a critical diagnosis of our time, a moral reflection, an essai in the best French sense of the word. The moral, if you will, of the essay is revealed in the title of its conclusion, 'ne pas se tromper de dépassement': 'do not be mistaken about overcoming', translated here as 'What "overcoming" means'. The context here is the contemporary questioning of human identity, and the prospect – greeted by some as utopia and by others as dystopia – of an overcoming of the human in some sort of post-human condition. Signs of incipient post-humanism are everywhere: from the cultural fascination with the figure of the monstrous in Mary Shelley's *Frankenstein* and its myriad cinematic variants and descendants, through to the science fiction world of cyborgs and artificial intelligence, to apocalyptic interpretations of contemporary nanotechnologies, genetically modified *enfants à la carte* or just plain old Dolly the sheep from Edinburgh. A recent and particularly fatuous and influential version of the fantasy of the post-human can be found in Michel Houellebecq's *Atomised* which identifies the possibility of a post-human future through genetic manipulation. This is the theory of what Houellebecq calls 'metaphysical mutation', which also incidentally entails the elimination of philosophy and the human sciences. Houellebecq writes, 'THE REVOLUTION WILL NOT BE MENTAL, BUT GENETIC'.[16] To those of us reared on the novels of Aldous Huxley, this is familiar fare, whether the dystopia of *Brave New World* or the utopia of *Island*. The question is: what is one to do faced with the prospect of the post-human?

Without ever retreating into an anti-scientific conservatism, Janicaud's counsel is clear: '. . . for the foreseeable future, it is not probable that the human being will cross the frontier and escape from its condition'.[17] He is equally firm in his opposition to the various forms of structuralist anti-humanism that emerged in the wake of the debate (or rather non-debate) between Sartre and Heidegger. Janicaud writes,

'Let us state it clearly: the indulgence that was shown in the 1960s for various utopias was fallacious'.[18] Thus, the claim for an overcoming of the human is a myth. Furthermore it is a myth that is complicit with a scientistic and deeply anti-philosophical conception of progress. As such, the claims for any sort of overcoming are a feature of what Janicaud in *Powers of the Rational* calls 'techno-discourse', with its basis in publicity-hungry scientists and inflated by the sensationalizing amnesia of the mass media. For Janicaud, what is morally problematic with aspirations towards the post-human is that they risk collapsing into the inhuman, whether it is the Bolshevik desire for the new man, the racial science of National Socialism or other variations on Ernst Jünger's category of 'the Titanesque'. The previous century was painfully replete with myths of the overcoming of humanity that legitimated the most inhuman of horrors. Janicaud writes, 'the utopia of an overcoming of the human is replete with inhumanity'.[19]

So, if the target of Janicaud's critical *Zeitdiagnose* is the fantasy of an overcoming of the human condition, and one perceives a clear analogy between this claim and his approach to metaphysics and rationality, then what prognosis follows from this diagnosis? Janicaud's view is more complex than might at first appear because the recurring fascination with myths of the post-human cannot simply be dismissed. But let's ask: if the humanity of the human cannot simply be overcome through an act of will or a new theory of metaphysical mutation, then what is the difference that characterizes the human? Summarizing his argument at the mid-point of *On the Human Condition*, Janicaud writes,

> We can return, then, to man and the human, not to revel in
> this issue, but to better understand the ambiguous richness

of a condition which no monstrousness allows us to evade. Man thinks he can leave his condition behind, whereas all these 'departures' only take him back to this fundamental truth. Humanity is the unfathomable overcoming of its limits.[20]

The seductive power of the various fantasies of overcoming is not just evidence of human stupidity. Rather, humanity itself might be defined by the restless attempt at the overcoming of its limits, the endless reshaping and reinvention of the human condition. The desire for overcoming is therefore a consequence of what, for Janicaud, is the most basic human characteristic: liberty. The dialectical paradox here is that the consequence of free human activity is subjugation to myths of the overcoming of the human condition that place in question that very freedom. We are free to err, it would seem. In his concluding paragraphs, Janicaud writes, 'A humanity that stopped wondering about itself would cease to be free'.[21] As I see it, a deeply Pascalian anthropology underlies Janicaud's argument in *On the Human Condition*. He continues:

Three hundred years ago, without having need of all our technological marvels to arrive at this intuition, the brilliant Pascal saw right through the irreducible ambiguity of the human condition, its instability and its balance between extremes (destitution, greatness), without sustaining the illusion of definitively warding off this always rekindled, sometimes unbearable, tension between the beast and the angel.[22]

We are divided between beast and angel, between an endless and endlessly frustrated desire for overcoming, for the post-human, and by the equally endless risk of falling back into the worst excesses of the inhuman. This situation is that of the human *partage*. Janicaud writes, once again turning to Pascal at a crucial moment in his argument,

> Indeed, the division (*partage*) between inhuman and superhuman actually corresponds to the two fronts on which man, this chronically unstable being, struggles to stabilize his existence: between inhuman regression and superhuman overcoming, between bestiality and angelism, between malevolence and deification. It should be emphasized that Pascal knew how to lay out the unstable and always surprising territory of the human, that "in-between" that results in "man infinitely surpassing man".[23]

The human being is this mortal and fragile *partage*, this division between the post-human and the inhuman, a *partage* that is also our lot, our allotment, the thrown contingency of our being. Otherwise said, the human being is a paradox: both beast and angel, divided against ourselves, defined by a conflict that constitutes us, but which is the very experience of our freedom, a freedom that constantly risks inverting itself into captivity. The human being is a movement of non-self-coincidence, a *partage* between what Max Scheler would see as the hiatus between *Sein* and *Haben*; between being and having, between the beastly material creature that one *is* and the angelic thoughtful reflection that we *have*. We *are* both *Sein* and *Haben*, that is, we are a paradox – as one might say in German '*ich bin, aber ich habe mich nicht*' ('I am, but I do not have myself'). The beastly and the angelic, the material and the spiritual, the physical and the metaphysical do not coincide, which means that we are eccentric creatures *par excellence*. We live beyond the limits set for us by nature by taking up a distance with respect to ourselves in the activity of free reflection, yet we are always caught in the nets of nature. We might even go so far as to say that the human being is the experience of this eccentricity with respect to itself, this hiatus between the beastly and the angelic, the inhuman and the post-human, the physical and the metaphysical, being and having.

*

Let me close by considering another central word in Janicaud's philosophical lexicon: *enigma*. Janicaud's thought is an activity of philosophical intelligence that moves between extremes: between instrumental rationality and the thinking of being, between metaphysics and its overcoming, between hyper-rationality and irrationality, between the post-human and the inhuman, between beast and angel. But it does this not in order to find a compromise or an Aristotelian measure between extremes, but as an act of fidelity to an enigma. The figure of enigma recurs in Janicaud's writing, most strikingly in the closing chapter of *Powers of the Rational*, where the very possibility that is envisaged as the *puissance* of the rational is revealed as 'the intelligence of the enigma'.[24] The movement of thought is here conceived as a response to the enigmatic, which is ultimately the enigma of our *partage*, our human lot, our fragile mortality, or, in the final words of *Powers of the Rational*, 'our future'.[25] Paradoxically, the phenomenological task consists in eliciting an enigma that resists any phenomenological description, the opaque gravity of human facticity. How then to understand the enigma of our being? Well, the point perhaps is not to understand it, but to elicit its features indirectly, however we may, through metaphors, dialogues, images, stories and the entire experimental activity of thinking. With this in mind, let me turn for a last time to Pascal, for *Powers of the Rational*, like *On the Human Condition* closes with an allusion to Pascal. However, a few pages earlier in *Powers of the Rational*, we read the following passage,

[. . .] Philosophical order is that of the 'Heart' (*le 'Coeur'*) in the specific sense that Pascal understood it, and where one cannot simply say that it is identical with the notion of feeling because it also maintains an essential relation to calculation. Thinking *as such* in the sense that we understand it is on the side of the Heart . . . a thinking that does not reduce itself to the fact that one thinks, or even that one thinks with exactitude or virtuosity, but rather that one thinks thinking itself (in our terms, that one meditates upon the enigma that there *is* thinking). (. . . *qu'on pense la pensée même (en nos termes: qu'on médite l'Enigme qu'il y ait pensée)*).[26]

The enigma is ultimately that of the heart. It is the heart which, for Pascal, has its reasons of which reason knows nothing. It is the heart which cannot be reduced to rational explanation, but which obligates the exercise of rational thought. It is the heart that is the enigmatic movement of thinking as such. It is this heart that beats, and that will beat forever, when we read Janicaud's work.

*

A final confession: Janicaud was the director of my M.Phil thesis which was, unsurprisingly given the argument of this paper, on the question of the overcoming of metaphysics in Heidegger and Carnap, a topic that he assigned to me and carefully supervised.[27] During my year and a half in Nice in the mid-1980s, we met regularly and he would sit patiently as I explained some text of Hegel, Heidegger, Ravaisson or whomever in my demotic French. He was a good, kind and generous man, of great integrity, hospitality and warmth. He was intellectually and geographically remote from the paranoid and finally provincial world of Parisian philosophy and his life in the provinces paradoxically gave him the liberty of a much more international outlook than most other French philosophers of his generation. The first volume of *Heidegger en France*, the magisterial last work that appeared in his lifetime, is interspersed with fascinating autobiographical epilogues, where Janicaud recounts his philosophical history. The last of them concludes with the words, '*qui vivra verra*', '*who will live will see*'.[28] Sadly, Janicaud will not live to see the impact of his hugely impressive body of work.

There is now an unprecedented uncertainty about human identity. This uneasiness (and that is putting it mildly) is due to a widespread subversion. This subversion relates first to knowledge of the origins of man and his point of attachment to the chain of beings: neither his genetic code, nor the use of tools, nor a certain language, nor social codes differentiate him in an absolute manner.[1] The subversions that revolve round his future could prove to be more radical still, especially if, at some point in the future, genuine biotechnological mutations were to transform 'the human race' to the point of rendering it unrecognizable, biologically, technically, culturally.

But the most serious subversion is of a psychological order: man is beginning to doubt his ability to fulfil his own destiny. In view of what he has done to himself and his environment, can he retain confidence in his own abilities to make judgements and assume responsibility? We are emerging out of a century of war and extermination where all the rules of human conduct were trampled under foot, and we are entering an era of 'globalization', which speeds up the dissemination of technologies, the exchange of goods and information, but which dramatically increases inequalities, unleashing acts of violence without precedent. The solutions and assurances that were expected to result from lightning advances in science and technology (especially in bio-engineering, in the cognitive and computer sciences) are either lacking or inadequate: these advances open up so many possibilities of manipulation both below our consciousness (biologically) and beyond (psychologically) that they in turn seem to endanger the humanity of man. Are we on the brink of a qualitative leap that would make us overstep the limits of our mortal and human condition? The aim of this book is to reflect upon the sense (or non-sense) of this potential 'beyond'.

But what is it to be human? Only man can ask himself this question, without ever being certain that he can satisfactorily answer it. Those who would advocate overcoming (or displacing) the human are sure to take advantage of these formidable uncertainties: if the specificity of man is more and more difficult to define in purely biological terms, if this specificity also diminishes from a cognitive point of view, if the notion of 'human nature' is obsolete, are the boundaries between human beings and animals, between what man is now and potential humanoid mutants, not very fragile?

Yet, is this question of a possible overcoming of the human generally well formulated? To what does the word 'overcoming' correspond for those who do not assign any limit to the applications of science and technology when it comes to human beings? Are the humanists, alarmed by the acceleration of all sorts of experiments on man and the human, right to speak of 'peril' or 'danger'? Even before clarifying this point, must we not agree on 'the humanity of man'? The task appears overwhelming, but it may not be insurmountable. And chatter must not suppress it. It is a fact: the discourse on bioethics is growing even more rapidly than that concerning the problems of how to make 'good use' of the internet, of media networks or on-line data processing. This proliferation

– inevitable and generally legitimate – could produce a background noise that obscures essential philosophical and ethical questions that we should be able to analyse.

The reader, for her part, is eager to go to the heart of this questioning, ridding herself of all rhetoric and ideological bias. How right this is! To assent to her request, the author must renounce all complacency, any withdrawal into an overly technical vocabulary, making himself clear and concise: a healthy exercise!

It seemed to me to be necessary 'to clear the way', proceeding, in this spirit, to an initial inquiry into what is at stake in humanism. How are we to clearly envisage the prospect of a possible overcoming of the human if we do not agree on the meaning of the humanity of man? Of course, it is neither possible nor desirable to consider this question in terms of its entire history or in all of its philosophical implications.

I have elected to begin by analysing the debate on humanism, which ran for half a century and even quite recently had fresh consequences. In reconstructing it dispassionately and in demystifying it, I could give the impression that the debate was only a matter of a quarrel over words. The question has at least to be stated plainly. It will require us to isolate 'what resists' in this discussion that is undoubtedly over-burdened with ideological and partisan ulterior motives. We will make it as philosophical as possible, updating the debate.

Our topical book goes beyond any announcement about the astonishing technological advances in particularly sensitive areas: does not increasing man's powers by a considerable margin mean shifting the frontiers of the human? If we had asked a man of the Middle Ages the question: 'Is a being that flies from Paris to Rome in two hours still a man?' he probably would have replied: 'It is a particularly swift bird or an angel, but not a man'. If we ask ourselves the question: 'Is an electronically aided being that reproduces through cloning still a man?', would we not be tempted to answer in the negative too? In both cases, the actual capacities of man seem to define him. Now the evolution and history of the human type have shown that man is precisely the being who continually exceeds the frontiers of his field of action, sometimes to the point of being no longer recognizable or identifiable in his own eyes. But is it the superhuman, the inhuman or other expressions of the human that lie in wait for us? Are not the dangers of monstrousness, which haunt our humanity, only myths that are reserved for science fiction? Not exclusively. We must compare myths and fictions to reality, encountering terrifying regressions into the inhuman. Are these warded off or nurtured by the call of the superhuman? It is to this question that we should try to reply next, once again taking myth into consideration in the assessment of risks that could always border on the unforeseeable. Finally, it will be important to show the ambiguity of the call of the superhuman, between the nobility of the desire for transfiguration and the vertigo of reduced capacities.

In view of this gap between the superhuman and the inhuman and also quite a few dangers, some of which are real, others fake, how are we to avoid facing the problems of regulation, of control, indeed of prescriptions? We will try to put forward a twofold strategy, both defensive (towards the inhuman) and open (to what 'passes man' in man). Overcoming prejudices and purely ideological oppositions, this strategy will afford at least the advantage, and perhaps the originality, of expanding the horizon and of opening up this salutary prospect: not to lose hope in ourselves.

One

In view of the monsters that endanger us and in line with the foreseeable and unforeseeable risks just mentioned, is the appeal to man's humanity, that is to say, to humanism the only course left open to us today?[1]

This course of action is continually disputed. You can find an account of it in some well-known and fairly recent philosophical debates, and it is interesting to piece together what is at stake in them. So why, quite suddenly, in the 1960s, especially in France, did the question of man crystallize in the rather polemical terms of a critical reflection on humanism? Without being able to launch into a lengthy historical overview here, we shall recall an obvious fact about man's awareness of his own humanity and sketch a brief retrospective of the humanist debate, from the 1960s to the present.

The evidence is that man only raises the question about his 'nature', 'essence' or 'properties' following a maturation (of the organism), an evolution (of the species) and a history (of society). It is now known that pseudo-'human nature' is the product of a biological evolution of several millions of years, a psycho-socio-linguistic maturation of several hundreds of thousands of years, and a techno-historical development of several thousands of years. Anthropology only emerged as a science in the nineteenth century. Of course, man provided symbolic representations of himself through mythologies, and about two and a half thousand years ago Socrates set out the advice of the Delphic oracle in rational terms: 'Know thyself!' But how many realize that all this 'knowledge' was the fruit of painstaking efforts, that the human sciences are still quite young (especially in comparison with the vastness of biological evolution) and that, in the main, man remains an enigma in his own eyes?

The summary of this evidence (or this collection of significant statements) has to be supplemented with that interest in the human which was condensed into the term 'humanism'. If the word itself only begins to be used systematically in French in around 1850 in the work of Proudhon, if it is sanctioned in ordinary language by Littré only at the end of the nineteenth century, it is from then on – for over a century now – the focus of serious reflection and debate on the very meaning of the human. If one wanted to take a broad view, one would say that traditional and classical thought (until Kant poses the critical question: 'who is man?') conceived the essence (or the nature) of man within the context of an ontology that was organized into a hierarchy, either in a naturalistic manner (with a good many variations from Aristotle to Lucretius and Cicero) or in a theological manner (from the creation of the world and of man by God). However, as far back as Hellenic-Latin antiquity a 'humanist' sensibility emerged insofar as the educational concern driving the elite presupposed the regular study of *litterae humaniores*, that is to say, the programme of great-literature, guarantor of the formation of the complete and noble man that the Roman citizen or true aristocrat was supposed to be. It is this tradition that the Renaissance wanted to revive. In current usage, humanism in this last sense has faded into the background, but cannot be said to be unknown, even if the most general sense of the term corresponds to 'every theory or doctrine that takes the human person and his flowering as an end'.[2]

Faced with this definition and this tradition, might not a common-sense reaction consist in looking at humanism as something patently obvious? 'How can one not be humanist?' you will say to yourself, casting into the outer darkness of absolute evil the major and minor criminals who bring disgrace upon the human. Indeed, it should be noted that in the twentieth century just about every school of thought, materialist as well as spiritualist, claimed to be humanist. Even Stalinists and Nazis could be seen declaring themselves humanists! This apparent consensus could only devalue the very term 'humanism'; it was too often turned into a label, without real content or coherence and become something of a cliché: 'You ask . . . "How can some sense be restored to the word *humanism?*" Your question not only presupposes a desire to retain the word *humanism* but also contains an admission that this word has lost its meaning.'[3] Replying in this way to Jean Beaufret, Heidegger is not content with noting the devaluation of a word. After all, the history of language is made of such phenomena, of the weakening of certain terms to the advantage of others. But what lies behind this phenomenon? Heidegger seems pretty much to be arguing against humanism here. How is this possible?

To understand the anti-humanist arguments out of which the 1960 to 1970 French debate was formed, it is logical to begin from Heidegger's position, as set out in the *Letter on Humanism*. It caused a considerable stir in France at the time, given that it was a rejoinder to the thesis defended by Sartre in his celebrated lecture, *Existentialism is a Humanism*.

THE HEIDEGGERIAN CRITIQUE OF SARTRIAN HUMANISM

We move now to a reputedly very difficult type of philosophy. For all that, it seems to me that the *Letter on Humanism* provides relatively easy access to one of the major themes of later Heidegger. I want to take up the challenge of demonstrating that here as clearly and concisely as possible.

Heidegger and Sartre share an opposition to the classical conception of a 'human nature' as comprising fixed and absolutely defined qualities. It is precisely from Heidegger that Sartre borrowed the celebrated proposition that instituted existentialism: 'Existence precedes essence'. For both philosophers, no abstract definition is really suitable for the being that we are. 'Man is all the time outside of himself: it is in projecting and losing himself beyond himself that he makes man . . . exist.'[4]

Going even further, Sartre challenges facile humanism, which consists in looking at certain human achievements (technical, sporting, etc.) and proclaiming: 'Man is magnificent'. In this case it is a matter not only of a superficial and self-satisfied humanism, but also, and especially, of the human world closing in on itself, as in the case of the cult of humanity extolled by positivism. Sartre specifically states: 'The cult of humanity ends in Comtian humanism, shut-in upon itself, and – this must be said – in Fascism. We do not want a humanism like that'.[5]

So far we have not yet seen what it is that can distinguish Heidegger from Sartre: opposed to fixing the essence of man in a classifying definition, both turn away from a humanism which would make man into an object of admiration and self-worship. Man must not think that he is a supreme end.

So, from this common starting point, how are we to understand Heidegger's declaration that he is radically critical of

Sartre? For what does he reproach him? For taking up human-ism in his own name, in spite of everything, without assessing what it really presupposes and represents. But does Sartre speak without due thought in proclaiming himself 'human-ist'? Certainly not: he upholds a philosophical position that is coherent and quite in conformity with Western tradition (we find it again in Marx): man is what he makes of him-self; becoming aware of his freedom, he must take on the responsibility for his future and work with a still more human world in mind: 'This is humanism, because we remind man that there is no legislator but himself'.[6] For Sartre, man was not created by God: thrown into the world without having requested it, he finds his way battling against the harshness of his situation and the absurdity of a condition that he has to share with other men. Hence, the exclusion of every possibil-ity of transcendent salvation: 'There is no universe except the human universe, the universe of human subjectivity'.[7]

It is this very closing of the horizon behind human subject-ivity that Heidegger radically challenges. If man withdraws into his own subjectivity, he refuses to be touched by the richness of what is; he turns away from the 'truth of Being'. Thinking himself master of things, he embarks upon the activism of a practice, forgetting his origins and his destin-ation. The crises and wars of the Western world are the dra-matic symptoms of this loss of the sense of a dwelling place, respect for the earth and a sense of the Sacred.

In naming man 'Dasein', in describing him as 'a creature of distance' and 'shepherd of being' – expressions that have not ceased to amaze – Heidegger wanted to take as far as possible the refusal to classify man in terms of any essence; he wanted to think out and safeguard what radically differentiates man from all animals. If Sartre really had seen that man is a project and is always beyond himself, he made the mistake of relating this capacity to subjectivity; that is precisely how he remained a metaphysician, an heir to Descartes and his 'I think'.

Refraining from any complacency about inhumanity and barbarism, Heidegger provides a critique of humanism because he thinks that it is not equal to the complexity and the possibilities of that being (*être*) who thinks he knows himself in describing himself as 'man'. Heidegger invites man to wonder at himself and think out the radical *difference* that distinguishes him from every thing, every animal and every being ('*étant*'). You might think that by reminding man that he is not the source of his own existence, and nor did he create the world, Heidegger sends man back to God. Insofar as the entire critique of anthropocentrism, thus conducted, is supposed to be radically non-metaphysical, Heidegger places the most original term, 'Being' (i.e., first relationship) before thought. Recapturing this first relationship is, for him, a condition of every new approach to the Sacred.

What should we retain of this critique of humanism? It has the immense significance – in wanting to be more true than any other position to the *radical difference* that characterizes man – of giving completely new life to the rejection of all anthropocentrism; man must learn again to go out from himself, to understand himself from Being, and no longer solely from his own subjectivity. Yet, Heidegger only makes a success of this 'campaign' thanks to a not inconsiderable over-simplification of classical humanism: can we hold without qualification that the entire classical tradition – that complex chain from Hellenic-Latin antiquity through the Renaissance continuing to the concern for 'honesty' (in the seventeenth century) and the Enlightenment (in the eighteenth century) – is only a metaphysical fixation, which forgets and misjudges the most different within that which makes human difference itself?

There was already an overstatement in Sartre whose attitude, at its core, consisted in defying the traditional humanists: 'I am more humanist than you!' And Heidegger, as it turns out, seems to say against Sartre: 'I assess the very *significance* of humanism still more radically than you!' What are we to think of these overstatements of the case?[8]

In demonstrating that they are largely misunderstandings (because they make the issues truly at stake less clear) I do not want to underrate the significance of these critiques: man must not revel either in the inventory of his 'qualities' nor in his achievements; his freedom is unfathomable; he can be the author of the best and the worst; he has to reposition his being in relation to what caused him to emerge in the world and in relation to the life that supports him and whose sense he bears.

Is it sufficient to be a humanist in order to take the true measure of man and his paradoxes? Surprisingly, despite its good intentions, humanism harbours a danger of closure and perhaps even of taking things for granted. We have to re-think man's being. It is to this task that, from a totally different angle, Claude Lévi-Strauss invites us.

DOES STRUCTURALISM TAKE SIDES AGAINST MAN?

With regard to structuralism, we must first guard against a serious confusion between *method* and *doctrine* (or ideology). It is reasonable to look for the origin of structuralism in the works of Saussure, Dumézil and Lévi-Strauss. Saussure looks upon language as a system of signs, disregarding its contents; Dumézil isolates the functioning of the triadic system formed by the Sovereign, the Soldier and the Artisan in the mythologies and social structures of Indo-European populations; Lévi-Strauss brings his method into focus by studying the basic forms of kinship, according to their inner logic, and examining their structural exchanges in groups of human beings that differ sharply from one another.

In none of these cases does the structuralist method constitute either an ideology or a philosophy, strictly speaking. A scientific method is neutral; it does not have to take a stand either for or against the various forms of humanism. There is nothing new in this respect in relation to Durkheim's methodological statement: 'Social facts are things'. This rule of sociological method, objectivizing suicide, for example, as a 'social fact', in no way excludes the possibility of an assessment of suicide on another level – for instance, the ethical.

Yet, structuralism has become 'anti-humanist', or at least, has been received as such. However questionable the confusion between method and ideology might be, can it be sorted out? We may first note that this confusion was mainly the work of interpreters or journalists from outside the scientific practice itself. Around the kernel of the founders of the structuralist method, one could see the formation of an ever-changing ideological scene, fewer scientific practices (e.g., in the literary domain with Roland Barthes) and, highly paradoxically, a 'structuralist sensibility'. While Dumézil was quite reticent with regard to ideology, this was not the case with Lévi-Strauss whose well-founded and spectacular opposition to Sartre in the area of the philosophy of history powerfully sustained the controversy. His unequivocal attack on the theses of the *Critique of Dialectical Reason* ineluctably entailed ideological repercussions because the discussion extended beyond the strict territory of scientific method: in the final chapter of *The Savage Mind*, Lévi-Strauss also threw in the reading of Marx, the concept of history and dialectic, the practice – or critique – of a philosophical anthropology, etc.

'I believe the ultimate goal of the human sciences [. . .is] not to constitute, but to dissolve man', writes Lévi-Strauss.[9] Reading such a declaration, more than one reader got caught up in the game: taken as a whole, this is an anti-humanist thought! How could we have doubts about it given that it attacks Sartre's humanism, his vision (considered 'naïve') of history where there are men who act, fight, engage in revolution, etc., where their subjectivity is constantly appealed to, where brotherhood and terror are regarded as decisive factors in a 'hot' chronology, both dialectically and anthropomorphically articulated? In the same way, Dufrenne confuses structuralism and anti-humanism in his book *Pour l'homme* (*For Man*). With great intentions, but rather incautiously, he carries out a 'cross-country' defence of man, as though he were morally threatened by a methodological mutation in what are known as the human sciences.[10]

This unsubtle defence of humanism is harmful, and adds to the confusion. In actual fact, Lévi-Strauss is not in the least anti-humanist in the practical sense. He initiated a distinction (perhaps too subtle for some people) between a scientific (or theoretical) method, which dissolves man as epistemological object, and a philosophical thinking, which re-introduces him as unity, value and sensitivity itself. This distinction will become a *distinguo*, clear but somewhat scholastic, in Althusser's neo-Marxism.[11] And it will not be absent from Foucault's works, although it will be inscribed more subtly there.

With Lévi-Strauss, we would have to have the time to re-read *Tristes Tropiques* in order to show when the ambivalence in the text between the cold gaze of an ethnologist who likes to think he is pragmatic, and the concession of a sensitive philosopher, an aesthete, moved by the sight of human tenderness in the Amazonian Indians amongst whom he has lived, first emerges. 'To study man one must learn to look from afar': Lévi-Strauss intends to put this piece of advice from Rousseau into practice in the field of scientific ethnology, where the ideological unity of the idea of man could not be presupposed without putting an obstacle in the way of research.[12]

What is more surprising is that Lévi-Strauss reintroduces humanism in the very practice of ethnological science, as can be seen in the field in *Tristes Tropiques* and in the most revealing text in this regard, 'The three humanisms'.[13]

Through its systematic interest in man, is not ethnology the human science *par excellence*? Relying on this premise, Lévi-Strauss can affirm, 'Ethnology . . . is the most ancient, most general form of what we designate by the name of humanism'.[14] The three humanisms that he distinguishes – classical, exotic and positive – are linked to ethnological practices: classical or traditional humanism (that of the Renaissance and the Jesuits) is, as a 'change of scenery technique', an unconsciousness ethnology; conversely, modern ethnology – turning to civilizations that are still looked down upon – practices a real humanism that does not necessarily declare itself to be one; situated between the two, 'exotic' humanism (of the seventeenth and eighteenth centuries) broadens out into new investigations, finding literary expression in, for example, Diderot and Rousseau.

A humanism that is connected to the very practice of a science can seem like a terribly scientist position. Yet, Lévi-Strauss's way of defending this thesis clearly presupposes a philosophical plan, faithful to the rationalist, enlightened and pro-democracy tradition.[15] But, however respectable it may be, this conclusion can occasion surprise inasmuch as it is a long way from the supposed 'anti-humanism' structuralist. We may judge as follows: 'By bringing together methods and techniques borrowed from all the sciences to serve in the understanding of man, it [ethnology] calls for the reconciliation of man and nature in a generalised humanism'.[16]

*

Although it seems to be inevitably on the defensive in relation to technical innovations, does humanism always rise from the ashes? The significance of the critique of its earlier or inadequate forms is that it forces us to think about man in a more demanding way: as free project in Sartre, and as *Dasein*, bearer of the very clearing of being in Heidegger. As for Lévi-Strauss, his noticeable about-turn (from a theoretical anti-humanism to a practical humanism) is explained by the purificatory task that he assigns to ethnological work, extending beyond traditional humanism both from the outside (the recourse to physical anthropology, prehistory, technology) and from the inside (the life of the ethnologist in the midst of a group with whom he empathizes).

In Foucault we again find – making due allowance for difference – a comparable gap between a theoretical anti-humanism and an emancipatory practice. Who more often than Foucault proclaimed the 'death of man'? His great book, *The Order of Things*, gave this theme an impact as extensive as it was enigmatic: this 'death', presented as the other side of the death of God, seemed to many to be even more mythical than the latter. What exactly did it mean? Foucault did everything to intrigue in advancing this theme; and he perfectly achieved his aim. Yet, strictly speaking, the 'death of man' proposition, which has nothing empirical about it, and which, in addition, has not yet been accomplished, is only the announcement of an inevitable epistemologico-philosophical mutation: the end of the two-hundred-year-old unified and sovereign figure of knowledge. Since the empirico-transcendental intensification, effected with the Kantian moment, man has occupied 'the place of the king', the privileged – but ambivalent – position of the self-foundation of a theoretical gaze whose positive object is none other than itself.[17] It is this reflexive, fragile and rather confused unification, which Foucault calls 'the anthropological sleep', that shows its fragility today.[18] In short, and to express it more plainly, it is the illegitimate side of what are known as the 'human' sciences that is poised to appear, without pretence, allowing the entire anthropologico-humanist vocabulary that accompanied these sciences to collapse. What is actually happening is the advance of positive knowledge (in biology, linguistics and economics, in particular). And standing out on the horizon is the 'return of masks', predicted by Nietzsche; that is to say, the breaking up of the substantial subject (the model of which is the Cartesian *cogito*) into the multiplicity of systems according to which life, signs and exchanges will be arranged in the future.

Yet, in his practical work, Foucault has campaigned against repressions and oppressions, and for the rights of prisoners, homosexuals and minorities. Even if it is not a question of issuing him with the signature to a blank moral document (for which moreover he has never asked), it is difficult to dispute the claim that his philosophical activity is in line with the emancipatory inspiration of the Enlightenment. Where humanism is concerned, must we not assess the significance of theories through reference to the facts? History's most terrible anti-humanists were called Stalin, Hitler, Pol Pot – and certainly not Althusser or Foucault. If the 'death of man' remains on theoretical ground, is it not easy to bid it farewell?

After all, we have yet to establish that there is a link between the theoretical and the practical – something the structural methods were not themselves able to ensure.

Thus, the critical discussion of humanism is a false debate only if, remaining rhetorical or excessively ideological, it conceals the fate truly reserved for humanity according to one or other ethical or political choice. It can and must allow a critique of its most conventional forms, a refinement of its senses and a distillation of its requirements. When all is said and done we debate and struggle over the limits of humanism on behalf of man, for an opening up of his horizon and for a liberation of his possibilities.

But is this liberation of possibilities, this general loss of constraints under the influence of scientific progress and the rise of technology, not fraught with new dangers? Is the humanity of man not immediately subverted by 'progress' and 'overcomings' that could well prove to be – especially in the ultra-sensitive domain of our genetic capital and our biological identity – sordid regressions or even the invention of monsters never before seen? The theoretical debates, then, could seem to be very academic. And it is to Peter Sloterdijk's credit that he has shaken our preconceptions and the conventional rhetoric of a good many of the discourses in bioethics, emphasizing when the techniques of genetic engineering and biotechnologies are in the process of completely changing the situation, that is to say the production and manipulation of human life.[19] Not only is our reproduction no longer surrounded by ancestral taboos, but the innovations (in vitro fertilization, surrogate mothers, the sale of ova, prospects of cloning, etc.) are such that a complete programming of human life by a politico-medical biopower no longer belongs to the order of science fiction.[20] Should we, for all that, consign all humanism to the rank of past and 'literary' inessentials? Should we resign ourselves to the opening of a 'human park', released herewith from all ethical courses of action? Are we ready, through complacency or indifference, to let a monster settle among us?

Two

Gabriel Marcel's unjustly neglected book, *Man Against Mass Society*, recalls a fundamental truth: men – capable of the worst as well as the best – are not only far from being angels, but also show themselves to be the most formidable enemies of the ideal of humanity.[1] The inhuman should not be sought beyond man. Nature, in its gifts as well as its disasters, contrasts us only with indifference. We either like or loathe animals: must they be strictly labelled inhuman? Is not a cat's stare radically *other*? We have to make up our minds to acknowledge – however painful this acknowledgement is – that man is, of all creatures, the only one who, through his violence, his barbarism and his sadism, can really show himself to be inhuman to the point of heinousness, precisely because he also possesses the possibility of being human. Man is the only animal who tortures and knows how to refine cruelty. Inhumanity is the other side of our coin, which proudly bears the ideal of the dignity and the greatness of man.

It is too easy to reject the inhuman outside of us: the very old and recurring temptation of the *alibi* which easily takes the form of monstrousness: dragons, the Hydra and Cerberus, fabulous and destructive monsters of mythologies, the moral monster (Satan and his demons), the political monster who terrorizes (from Attila to Hitler). And for some time now, we have seen – around the figure of man – new monsters lurk, virtual productions of science and technology. It is on these dangers that we shall focus our attention, once we have made an effort to clarify the actual notion of the monster.

THE MONSTER

Even before the imagination is angered or panics before the harrowing strangeness of its own fictions, life produces monsters. Let us clarify the biological definition of the monster in order to better understand what it involves. This definition will not help to draw an impenetrable barrier between the biological and everything else. Rather, it will allow us to discover the interweaving of life and games of make-believe.

The monster is 'an organism with an unusual configuration'.[2] This is a medical or biological concept (neither inorganic nor mechanical), which does not entail any value judgement, any shrinking back from the supernatural or the ghastly, as, on the contrary, was the case with the original Latin *monstrum*, a sign or warning of the gods. While the monster in the scientific sense represents a maximal divergence with regard to normality (an exceptional divergence, because if nature multiplied the number of monsters, they would then become the norm . . .) and is thus neutrally definable in statistical terms, things are seen quite differently from the moral point of view: it is not the fact that the moral monster is exceptional that satisfies the characterization; we loathe it because of its crimes, its appalling cruelty and its 'inhumanity'.[3] What does this mean? The corruption of a free being, the surrender to evil for the sake of evil. What does humanity then become? It degenerates, as if seized by a sort of regressive fever. That is why moral monstrousness provokes indignation, revolt, whilst physical monstrousness arouses only repulsion, pity or indifference. Moral monstrousness is an insane squandering of the possibilities of freedom. The fact that it is reprehensible does not imply that it is explicable: more often than not acts of extreme inhumanity (rapes, tortures, sadistic murders) leave us perplexed or incredulous, before provoking indignation. How, for example, can the murders of women and innocent children, as happened only recently in Algeria, be perpetrated in cold blood by men worthy of the name? Incensed, we also remain perplexed: we would like to be struck by the impossibility of such disturbing acts, we would like to believe that those who committed them were not men. And, in fact, did the SS not follow a training programme for suppressing all their feelings of pity? Man is thus confronted with the inhuman: a moral monstrousness in which radical evil is embodied.

There is a mystery of the inhuman, sombre conjunction of horror and gloom. The monstrous lets us then perhaps catch sight of the abyss of barbarism and cruelty from which humanity slowly emerged – origins that make us shudder with disgust as though we were before a taboo we should like to forget, fascination for a sacred disfigured in the diabolical. Yet how can we not know that the sight of blood spilled in a massacre can excite the torturers, that the cries, the groans could well render them even more cruel? How can we maintain that, exposed through exceptional circumstances to extreme situations of war or vengeance, we ourselves could avoid such madness? One should like to throw a discreet veil over these 'realities' of the psychology of depths. This defence reaction, however legitimate it is, testifies to the difficulty that the civilized, morally responsible person experiences, in taking on the unfathomable part of inhumanity that our humanity also contains.

Nothing obliges us to make an assessment of this extreme

inhumanity or to 'map the terrain' (insofar as that is conceivable, considering the inventiveness of humans where evil is concerned). There are criminologists and psychiatrists for that. Only an unhealthy lassitude at work on our imagination could have us say: 'I too would have been able to carry out this intolerable cruelty; I see myself committing this heinous crime'. It is not in this way that we must take on inhumanity. And the distinction to observe is none other than the separation that has to be made between the clear understanding of realities (including the psychological) and the moral mastery of these realities – which is a matter for a responsible will. In other words, our humanity does not require that we take it on in extension or quantity, but in quality, with propriety – without, however, ignoring the hard realities.

Thus we find ourselves between two asymmetrical abysses. To choose the Good is not to be an angel. But having too much to do with the beast, we run the risk of no longer being in a suitable state from which to renounce it. Man is this middle course between the angel and the beast, this mixture of the two: Pascal had perceived how unstable this middle course is and how difficult it is to find equilibrium. And he saw how man (on account of his fallen nature?) is much more attracted by abysses than summits.

To take on the inhuman is not to approve of it. To know that there is a monster lying dormant in each one of us must render us more vigilant still.

However, the problem of the monster is not currently uniquely formulated in terms of morality. It can be pointed out that a new type of monster lies in wait for us, not the 'classical' moral monster (if we can speak in such terms), but a technological monster, the result of systematic manipulations of the human species.

Again it is necessary to take myth and reality into consideration. In this respect we will examine a first myth, already two centuries old, but still vivid in our imagination.

THE FRANKENSTEIN CASE

If reading Mary Shelley's book, *Frankenstein or the Modern Prometheus*, is enormously instructive, it is so, as we are about to discover, on account of its psychological and human profundity and not for the reason ordinarily put forward. Published in 1818, this work offers an extraordinary forewarning of science's fantastic powers to manipulate life and especially the dangers of these powers. Dr Victor Frankenstein, a brilliant and enthusiastic young scientist, is carried away by the idea of creating a living being in a laboratory; he succeeds in doing so, but the creature produced immediately escapes, and the scientist realizes too late that what he has made is an odious monster which horrifies him. Despite his remorse and his good intentions, Frankenstein will not be able to stop the series of crimes and calamities – knowingly caused by the monster – that will surround him, lead him to despair and finally destroy him.

Frankenstein himself is in no way presented as a diabolical being. On the contrary: clever, hard-working, well-meaning, all he did was give in to a powerful, near irrepressible, impulse that was virtually imposed upon him by the scientific spirit itself. An apprentice sorcerer having carried out something irreparable, he is aware of his responsibilities, admits in his heart of hearts to being the real culprit, does all in his power to stop the inevitable course of events (he even refuses to honour the promise made to the monster to create a companion for him) – but is too late.

The fact that this work acquired mythical status (thanks

mainly to its countless cinematic adaptations) is not simply due to chance: why does the figure of Frankenstein fascinate us? It is because the Promethean dream of the creative power of human science over life is concentrated in it. Almost two centuries after Mary Shelley, we assess at what point this dream is becoming reality – and nothing guarantees us that it will not turn into a nightmare of even greater proportions than the tragic fate of Dr Frankenstein.

It is not a question of downplaying the truth of the idea to which Mary Shelley has given a disquieting and unforgettable force. To suggest that this idea is not the only one but that the work is still richer does not in any way amount to diverting attention from the brilliantly thought-out updating of an intuition already set down in the Bible: wanting to taste the forbidden fruit of the tree of knowledge of Good and Evil, Adam and Eve instituted our mortal and fallible condition. A contradictory and malignant creation, original sin renders all forms of evil possible. Yet, among this and as if secretly sustaining its proliferation, is there not the desire for a knowledge-based power? If this temptation of a dominating knowledge is not the only source of the sin, does it not consti-tute its most secret and fearsome motivation, the rarest, the most cunning, and the most diabolical? From Saint Augustine to Pascal, from the alchemists to Goethe's *Faust*, does the Western tradition not ponder over the immense dangers har-boured by the admirable and initially noble feats of science? Rising well above the concupiscence of the flesh and the satis-factions obtained through the direct domination of things and beings, conquering science strives to have us overcome our imperfections and our finitude: seeking the elixir of life, it aims to increase the power of our weapons, the speed of our vehicles; soon it will make us fly (even Leonardo da Vinci gave technical form to the ancient dream of Icarus). But traditional Christian thought made do with condemnation of an arrogant science which it believed was losing all respect for the Creator and the order of his creation; it preferred, with Saint Thomas, to rely on reason enlightened by faith: could it have been discerning (even at the time of Galileo's trial) what was beginning to take root in the desire to know? Without any theological or philosophical pretension, Mary Shelley's novel does more than express a premonition, because it dramatizes the nightmare to which the wild manipulation of life leads.

There is, however, a still more interesting feature in *Frankenstein*: the monster (who remains without a name) enters into dialogue with his 'creator'. And in spite of the repugnant character of his physique, he proves to be intelligent, aware of his situation, suffering due to his immense solitude, longing for kindness and sympathy. He is not in the least lacking moral sense. Yet will he not commit the most odious crimes? Mary Shelley, by letting him speak at length in the most striking part of the book, gives moving and human accents to his confession that do not fail to weaken the hostility of Victor Frankenstein and, through him, the horrified resistance of the reader. The monster explains that – not in any way responsible for his coming into the world, and still less for his repellent appearance – he had initially done his best to attract, for want of sympathy, at least the commiseration of humans. It is their inhumanity, hatred and the ill-treatments he suffers that pushed him back towards an intolerable solitude and an unshakeable thirst for vengeance: 'I am malicious because I am miserable. Am I not shunned and hated by all mankind?'[4] Created by a man, the monster finds himself rejected by all men and even by Frankenstein himself: his malice is almost imposed on him. And the image of the companion, whom he begs from Frankenstein, symbolizes a legitimate desire for love. 'Oh! My creator, make me happy; let me feel gratitude towards you for one benefit! Let me see that I excite the sympathy of some existing thing . . ..'[5] Does he not then embody both the desire for social harmony and – despite himself – that part of strangeness and horror that man rejects, incapable of acknowledging the full extent of his condition?

Is the monster created by Frankenstein beyond the human? Not in the least. Endowed with a superhuman physical strength, he commits inhuman acts, but he explains them and justifies them to the very end of the novel, demonstrating an admittedly 'monstrous' intelligence and sensibility – but always approaching the human. In desire and for want of love, he is led to despair. His final escape into the wastes of the far North projects the vertigo of the inhuman into the imaginary.

Frankenstein is the bearer of a double lesson: one, the more obvious, on the potentialities and the dangers of a science without conscience; the other, more hidden, on the psychological and moral risk that lies in wait for us: an inhumanity aggravated to the point of monstrousness. Science and technology undoubtedly bring about significant modifications of our powers. For all that, need we expect in these modifications mutations that go beyond the human? This leap into the imaginary carried out principally by science fiction (and which benefits from so much of our complicity) could well play the role of an excuse for the abdication of all responsibility. The pervading technologism leads us to think that we are on the brink of a great leap forward that will cause us to 'mutate'. It would become so easy then (and nearly 'automatic') to make objection to every reference to humanity, its weaknesses as well as its merits: 'All of that is overcome! We are in a new era! An entirely different race is born!' Are we not already acquainted with this theme, twisted premonitions of which the totalitarianisms of the twentieth century have offered? In light of the most recent developments in computer science and on-line data processing, one can imagine a more subtle scenario: the overcoming of the human by a Successor without face or body, but infinitely more intelligent and robust than us.

A TOTALLY INHUMAN SUCCESSOR?

Let us now examine a hypothesis recently developed by Jean-Michel Truong.[6] Does the author, a specialist in artificial intelligence and a novelist, merely add another stone to the already impressive edifice of science fiction? Not quite. He argues from incontestable data; and the future that he envisages is not at all improbable, given a sufficiently long-term view.

It is a matter of going well beyond a humanity improved or transformed by electronic, transgenic or 'bionic' technologies. Placing ourselves within the audacious perspective of an evolution that has not had its final word on the human type, and that proceeds via creative experimentation in every possible medium, we might dare to think that artificial intelligence initiates a quite autonomous operation. Robots capable of constructing other robots have already been created. With the internet, we shall see self-producing software multiply and the colossal work of 'e-genes' should produce a 'totally inhuman' intelligence (in the sense that we understand humanity today), passing – after millions of attempts and a complexification hardly conceivable today – to a new form of life, surviving an entropy that invisibly eats into our organs and our biological cycle, on a planet itself condemned in the long run to lose the light and warmth of the sun. The Successor would be, in a still undetermined future, a new species, a sort of post-humanity, entirely different from us.[7] We witness its emergence only from computer memories and the 'global interconnections' of the internet.[8]

There are two arguments for this hypothesis: the first relies on the unpredictable character of an immense evolutionary process, of which present-day humanity (which no longer manages to think about itself in terms of the concept of 'human nature') might only be a transitional phase:[9] does Bergson not see in the universe a 'machine for making gods?' And before him, did Nietzsche not predict the advent of the Superman? The second argument can rely on the fact that the globalization of electronic communications produces autonomic phenomena that are still in their early stages: the progress of computing and its universal diffusion were so incredibly rapid, the explosion of the virtual was of a kind that one can no longer argue in the classical terms of a linear development. The qualitative thresholds were crossed and we are already well beyond the problem of the mastery of technical instruments by a supposedly sovereign man. Humanity is dragged down by the 'logic' of the networks on which it now depends. Unpredictable evolution, new and specific constraints: these are two indisputable points.

On the other hand, what poses a great difficulty in the spectacular hypothesis that we are examining is the use of the very term 'life' and of the analogy of its evolution. Because if the support of the Successor is no longer carbon, it no longer has anything in common with life as we know it on earth. Let us suppose that it were an inorganic substance: silicon, for example. Again let us assume that the electricity necessary for the functioning of this vast network of networks were self-produced. Again let us grant many other technological 'leaps'. In the strict sense, the intelligence that would survive and even develop then would present the qualities and faults of a vast bank of self-programmed data, without any anchorage in flesh and blood. That being the case, why continue to speak of a new form of life? Why even describe this reality as 'totally inhuman' (that which still makes reference to the human)? Why call it 'Successor', as if the singular and still personalized framework of a 'succession' still made sense at this level?

If Jean-Michel Truong is right (and he is not entirely wrong), his remarks amount to acknowledging that there will always be, beyond man, physical and informational exchanges and perhaps even (using Teilhard de Chardin's term) a form of 'noosphere', that is to say an area of supra-material inter-communications, inconceivable within the limits of our incarnate intelligence.

There is inhumanity only for man and in reference to the idea that he constructs of his own humanity. The current operation of 'Internet-Globalization' can be largely inhuman, transgressing the limits and moorings that made man's 'humus' as we know it. That cannot be denied. But, to be clear, it is always man who exchanges all these pieces of information and who above all protests, rebels, worries himself about no longer recognizing either his face or his marks in the evolution that carries him along. Even though we speak of a 'totally inhuman' reality, the adverb 'totally' does not manage to erase the reference to the human. That does not mean that there neither is nor ever could be a reality that is totally *other* (what do we know of what 'takes place' on the fringes of the universe?), but it always testifies to the speci-ficity of human consciousness: human consciousness poses in opposing, and asserts itself in overcoming itself; in this sense, it cannot entirely extricate itself from itself, it cannot cut itself off completely from the remarkable relation that binds it to things and to itself – this irreplaceable bond is called 'subjectivity' or 'openness to Being'. That is to say that if the disappearance of the human is not impossible, the content of this disappearance remains as inconceivable to man as that which awaits him (or does not await him) beyond his physical disappearance: if I decide to commit suicide (*me donner la mort*), I do not really know what I 'give' ('*donne*') myself: I know only what I reject. What is true for the individual is not less so for humanity as a whole.

*

We can return, then, to man and the human, not to overstate the case, but to better understand the ambiguous richness of a condition which no monstrousness allows us to evade. Man thinks he can leave his condition behind, whereas all of these 'departures' only take him back to this fundamental truth: humanity is the unfathomable overcoming of its limits. And it is never shielded from the inhuman. It is as if, ever since the departure from the Garden of Eden, the vastness of the bare land also opened out onto a desolate infinity. It is this inhumanity that constantly lies in wait for the human, and it is to this figure of the monster, haunting the beauty of every man's and woman's face, that we have to return and consider anew.

THE CYBORG: MYTH TODAY, REALITY TOMORROW?

Frederik Pohl's science fiction novel, *Man Plus*, recounts the first space expedition by human beings to Mars.[10] Its interest for us stems from the fact that the astronaut who is its hero, Roger Torraway, is (or rather becomes) a Cyborg. Many months before his departure, he undergoes a series of surgical operations and implantations that completely transform his brain, his senses, his muscle structure and his skin, in order to adapt him to life on the 'red planet', but deprive him of his old human form. From now on devoid of sex, deprived of a face worthy of the name, equipped with ailerons, he is wholly assisted and controlled by batteries of computers. Following lengthy training, these transformations offer considerable advantages in terms of resistance, speed, night vision, etc.: this hero even becomes capable of learning, at lightening speed, to play the guitar. On Mars, he will have an exemplary agility and endurance, but neither the look in his eye nor his voice have anything human about them, and his reactions are predictable only in accordance with his programming (thus, when he arrives on Mars he perceives a gesture made by one of his companions as a serious threat and injures him, because of a defective technical setting).

The Cyborg is therefore a mixed being: man and robot all at once. The transformations that were imposed on him go beyond simple prostheses. The title Man Plus reflects a behaviourist conception that considers only the growth in performance. But is humanity merely a platform on which one could freely and unrestrictedly undertake technological transformations? Is humanity separable from the body that gives it its age-old configuration, and not simply a passing 'hospitality'? The novelist does not avoid the question, because he lets an astronaut who has to be the Cyborg's companion know: 'Colonel Roger Torraway, [is a] human being . . . As human as you are, except for some improvements'.[11]

After various incidents in the course of which the Cyborg goes wrong and becomes virtually insane, everything is resolved: the Cyborg is rescued and repaired, life is organized on Mars, other colonists get ready: 'We had saved our race [. . .]. The future of machine intelligence was therefore assured'.[12]

Without overestimating the importance of this novel, among a good many others of the same genre we can point out two significant features: the conquest of space is conceived as a chance for a humanity condemned to a limited time on a planet, to whose destruction it has largely contributed; the Cyborg, a very fragile mixed creature, is not presented here as a superhuman being or absolutely different from the human, but rather as a variant of the human – displacement and readjustment rather than complete transformation.

Now, on these two points, science fiction directly sustains contemporary concerns. It reflects our fears and serves only to urge on an already powerfully committed momentum: the conquest of space itself linked to the development of all the technologies meant to insure the survival of man in weightlessness or in extreme conditions, and the combination to this end of cybernetics and the biotechnologies. That the colonization of Mars by humans is not very probable matters less than the symbolic confirmation, strong and repeated, of an application of technological advances to the human organism itself, both in the operation of its nervous system and in its behavioural possibilities.

Condensed in the mythical figure of the Cyborg is the real dynamism of research already begun, which will perhaps result, in a manner which is still unforeseeable today, in 'improvements' comparable with those of which science fiction dreams. It could well be that bioengineering and nanotechnologies, combined with new advances in miniaturized computing, allow for the perfection (at a time perhaps remote from us, but 'reasonable' on the scale of the evolution of our species) of extremely robust, almost immortal, 'human specimens' endowed with a prodigious memory, an extremely quick and strategic intelligence, as well as a heightened

On the Human Condition

sensory palette. There is no doubt that our techno-scientific civilization is going in that direction, and is aiming at this type of goal, and nurtures this myth, because the junction between neurons in the brain and silicon chips has already been carried out experimentally.[13]

To that extent, it can be assumed that the conditions for the Cyborg are being put in place, and that it will ineluctably be conceived and built. Do we have to go as far as perceiving here the Superhuman in the Nietzschean sense? Certainly not, because a Superman with fabulous technical achievements would only be a caricature of Nietzsche's ideal. Does the Cyborg prefigure an overcoming of the human? Nothing is less certain: in order to assert it we would have to be able to 'classify' the humanity of the human and reify this irreducible kernel that still stands firm in us, and this would only be possible in terms of the power of self-transcendence. That being the case, how are we to think about a self-transformed humanity, both on its mythical side and in taking account of the inevitable limits of its projections into the future?

Let us sketch out an answer. The myth of the Cyborg tells us more about our aspirations (and our fears) than about the actual form that new increases in human capacities will take when applied to man himself. Whatever this 'progress' may be (and there certainly will be some), there is no guarantee that it must allow man to transcend his condition. The aspirations that it fetishizes in technology (immortality, invulnerability, sovereign intelligence) are as old as humanity itself. But instead of them being projected into a divine sphere or concentrated in magical forces, they are (we believe) at our door. In view of what has been made possible thanks to wonderful technologies (rapid travel, instant transmission of news and telephone calls, notable increase in life expectancy in developing countries, etc.), it is difficult to assume that the movement is not continuing.

At the same time, man works hard to acknowledge a fact that must be staring him in the face: on the essentials, that is to say his condition of being conscious, free, embodied, finite, on his responsibility to choose between good and evil, on that which serves as the heart of his situation, nothing has changed. He is always that unstable and fragile being that has difficulty accepting his margin of liberty, living amongst his fellow men and creating a wisdom for himself. Even in that novel with such a revealing title, *Man Plus*, we detect this resignation in the author (and his hero): the choice to live on Mars is still a human decision and the colonizers of the 'red planet' will always be fundamentally humans. However much man throws his body sidereal distances from the earth, he does not succeed – whatever he does to increase his powers – in escaping from his condition.

It is really quite surprising that the Cyborg is not more monstrous. What is really monstrous is the inhuman in the moral sense, evil willed for the sake of evil. The Cyborg remains very human, like a big toy. You can still play with it. But humanity does not let itself dissolve so easily. However inhuman the universe produced by technology is, it still refers to the human, which is its source, uniquely capable of using it and giving it meaning. That which endangers humanity, then, really derives from itself: a freedom that turns against itself. It is we who have created and nurtured this 'inordinately enlarged body' – Bergson's description of the technologized world – even if we no longer know how to live in it or to animate it. And above all, evil, always latent, always ready to spoil our achievements irreparably, is as old as man.

Three

In jumping from a discussion of the dangers of the monstrous to an assessment of foreseeable risks, are we not abandoning the mythical element in order to go back over the solid ground of facts? It is not as simple as that. On the one hand, the contemporary situation is not stable, but is changing constantly under the pressure of techno-scientific innovation. On the other hand, we have to take account of the omnipresent penetration of a 'techno-discourse' with a deliberately mythicizing component that contributes towards blurring the limits between the actual and the desirable, in support of politico-economic inducements, power struggles between laboratories, market dominance by powerful companies, and even trade promotions or 'brainwashing' in the advertising sense.[1] In short, this 'techno-discourse' is all the psycho-ideological propaganda that sustains, accompanies and promotes technological innovations.

If the dangers previously mentioned are not in any way to be brushed aside, it remains essential to have a more measured and at least long-term vision. At the risk of appearing almost too reasonable, this investigation will find itself compelled to 'make allowances' in areas that are quite sensitive at present. If we did not devote ourselves to this particular effort of evaluation, it would be quite right to reproach us for jumping directly from mythical dangers – too easily exploit-able in the confusion – to hasty conclusions, bypassing the most reputable data pertaining to very real problems.

We shall concentrate our efforts at evaluation on the question of cloning, a technique that became a reality on 5 July 1996 with the birth of Dolly the sheep at Edinburgh, and whose principle is still eagerly debated today. Without going into overly technical details, cloning is the reproduction of the same organism by transfer into the previously enucleated oocyte of an ordinary cell, taken from a fully grown organism.

The gap between myth and reality is evident here. From a single successful result in the animal domain (followed, it is true, by about a hundred others on cattle, sheep, goats and mice), the imagination gave itself free rein. Are we moving towards a reproductive cloning either for the purpose of eugenics or in order to satisfy fantasies of immortality?[2] Could the combination of the prospect of gain (on the part of laboratories) and the desire for power (on the part of wealthy individuals and totalitarian states) lead to nightmarish clon-ing, exempt from all ethical control, for example the mass production of clones for military purposes? In this particular case, it cannot be said that the imagination is totally 'wrong', because it plays a part in formulating an extremely serious problem: the potential danger that cloning could strain the very principle of the individual singularity of human beings. However, to return to the facts, cloning still falls far short of the target: the failure rate of present-day cloning techniques is important, the risks of malformations are not negligible, and the psychological opposition and the legal restrictions in place hardly contribute towards speeding up the process – nevertheless, it is unquestionably underway.

In the main the most fundamental ethical objections are directed at reproductive cloning. These objections are, of course, chiefly of a religious nature (man does not have the right to tamper with the very source of life to the point of cloning himself); but they can be replaced by a philosophical argument: whereas a man is an absolutely singular being, the potential clone will only be a copy of an earlier being; this identical copy undermines a human specificity: the singularity of each individual. The *enfant à la carte* will be no unique being, something that prompts Axel Kahn to protest:

> With what right and on what pretext are we to accept that men find themselves recognising this unique and unprecedented privilege of deciding that others are going to be born who will resemble them so closely, and so many of whose essential characteristics they will have determined?[3]

Can the distinction between therapeutic and reproductive cloning enable us to halt these objections? Henri Atlan has recently warned against the very expression 'therapeutic cloning'. In order to avoid all confusion, he is happier talking about a stem cell culture or the 'transfer of a somatic nucleus'. Whilst he is opposed to reproductive cloning, Atlan considers research on cells produced without fertilization and having no reproductive purpose to be legitimate.[4] Indeed, the proponents of stem cell cultures claim that an immense field would be opened up for research both in the area of therapeutic regeneration and in biological 'engineering', which would be infinitely more effective than the prostheses and transplants used now.[5] They also draw support from the fact that this research has already got underway in the United States and in Great Britain (where the ban on all reproductive cloning is clear and there are no grounds for transgressing it, at least for the time being).[6] Nevertheless, the mistrust of therapeutic cloning is fuelled by the suspicion that it might only be a stage on the route to reproductive cloning[7] – already in great demand and heralded, not without immodesty, by the Italian physician Antinori.[8] At present reproductive cloning has been universally rejected – with the above exception – although it is unclear as to how 'piracy' in this area can be prevented.[9] Axel Kahn, resolutely opposed to this form of cloning, has no illusions: 'I am convinced that in the 21st century . . . it will be carried out'.[10]

As always, in an area of this kind, we have to distinguish between 'is' and 'ought'. There is no ethical validity in the argument that runs: 'That will happen anyway; you can do nothing about it'; otherwise, this pseudo-realism could go as far as claiming: 'there will always be crimes; it is pointless to condemn them'. This lack of comprehension of the specific validity of the 'ought to be' is also fallacious in practice, because the recognition of norms and the setting up of jurisdictions make a contribution towards limiting the uninhibited development of 'anything goes'. Are we contemplating here what humanity would become, and not only in the area of biotechnologies, if it suddenly changed to the law of the jungle, without any kind of safeguard?

Let us, however, consider this hypothesis for a moment. Would humanity go beyond the human? It is likely that a regression into what we call, for want of something better, inhumanity, would occur. Alas, wars, massacres, terrorism today constitute so many 'pockets' of inhumanity on a planet where morality and the law have a great deal of difficulty in establishing themselves. It is almost impossible to answer the awful question that has just been raised: humanity is inseparable from its hideous and reprehensible other side; but how far can regression go? We can, unfortunately, ascribe no limit to inhumanity.

Continuing these disenchanted reflections (that are intended to clarify, but do not in the least intend to justify the unjustifiable), we also ask whether reproductive cloning, practised on a grand scale and without any ethical supervision, would facilitate an overcoming of the human. It is noted that genetically identical twins find a psychological and moral individuality through the contingencies of their cultural history as well as a singular faculty of creation. It is likely that the same would apply to human clones. It is not impossible – at the risk of giving way to science fiction again, but this time in support of the idea of humanity – for a 'prohuman' revolt to take place in a situation of extreme dereliction, wild permissiveness or regressive violence (the fight to the death between men threatened with destruction or enslavement and their torturers – highly sophisticated androids or robots – whom they themselves created is a recurrent theme of science fiction).[11] This means that the core of man's humanity is not necessarily dependent upon the 'centring' or the maintenance of one of its traditional characters: the spirit is certainly more mysterious and more paradoxical.

We should not, however, push this type of reasoning to the limit. To the extent that we reach the outermost limits of the very humanity of man in this manner, it is risky to make the slightest prediction. Nothing guarantees that reproductive cloning would lead to the depths of inhumanity; nothing guarantees the contrary either. But the acknowledgement of the fragility of man's humanity can only encourage a cautious attitude, applying the precaution principle each time that we broach the most sensitive subjects where the future of our species is scorned.

Could man's humanity fragment under extreme pressure? While recognizing its paradoxical character and its still possible revival in situations of extreme alienation, we must also consider its fragility. Man's humanity thus reaches its limits and borders on the unimaginable. After all, man's existence – even biological – is dependent upon physical and environmental conditions (oxygen, water, ozone layer, food and energy resources, etc.) that we know are finite and under threat, in space as well as in time. Human equilibrium in cultural terms (in the broad sense) proves to be equally unstable and fragile: the death of civilizations – even if this has become a very banal theme since Valéry – is well and truly a historical reality. Beyond the rhetoric, there are dreadful facts to take into consideration: the apparently refined civilization of early twentieth-century Europe did not prevent the most abominable mass slaughters in history; nothing allows us, at present, to rule out the multiplication of conflicts, forms of manipulation and terrorism that fly in the face of every 'humanist' imagination.

The main difficulty that we must recognize then is the following: the assessment of risks (in the area of bio-technologies as in any other field where technology shifts the limits of human powers) is as essential as the possibility and requirement that we maintain confidence in the capacity of man to 'bounce back'. But neither does that minimal rationality and that persistent hope need to lead to delusion about the fact that the future of humanity can escape all calculation of risk.

In the future, there would not only be that epistemological destruction of man's face, dramatically put by Foucault, which makes us forget about the human, but there would also be the advent of (what Ernst Jünger terms) the *Titanic*, beyond all regulation of a human–inhuman 'park', perhaps also beyond all predictions.

Can we at least make out certain contours of this unforeseeable future that lies in wait for us: will man be *overcome* by technology or will he be capable of *overcoming himself* – perhaps, mythically?

FROM THE HUMAN PARK TO THE ADVENT OF THE TITANIC

Among the pictures of the unforeseeable, of what emerges beyond the human, the hypothesis formed by Peter Sloterdijk stands out: a society that thought it was humanist will find itself constrained – due to advances in biotechnologies – to change, organizing a control of man by man, that is to say, the selection of the 'best'. Stockbreeders and livestock: will this division of the human herd, anticipated by Plato in the *Politics*, become our destiny?

Why did Sloterdijk's lecture *Rules for the Human Park* cause such scandal?[12] Because it coolly considered setting up, on a grand scale, a biopolitics that brings to mind – particularly in Germany – the most deplorable memories: a selection of the human herd in accordance with the rules of an anthropotechnology, leading to the creation of a 'human zoo' or park. Handled in this way, humanity would be divided into stockbreeders and livestock: the fittest would be destined to take on actively the task of bioselection. What has caused offence, above all, is the cynical (or provocative) recourse to a terminology of taming and selection, throwing humanism on the scrap heap for an 'animalization' of the better part of future humanity. Is this a return to a Nietzscheanism of the most questionable type or – worse still – regression to a crypto-Nazi biologism?

As a matter of fact, Sloterdijk neither approves nor

condemns this possible evolution. Considering it both 'inevitable' and 'insurmountable', he is content to mention it, without hiding his face before its unforeseeably frightening or ethically reprehensible sides. His standpoint is neither prescriptive nor sanctimonious. His detached irony and his frank references to two thinkers, considered 'politically incorrect' in Germany since the end of the Second World War – Nietzsche and Heidegger – breaks with the dialogical rationality, the rather edifying philosophical discourse of the Frankfurt School, old and new (from Adorno to Habermas).

We can calmly make the distinction here between the acceptable and the unacceptable. What is acceptable, and even highly desirable, is the clear formulation of new problems in the use of a constantly developing bioengineering. Noting that this development cannot be effected without new rules is almost a truism. Man, who already manipulates the origins of his own life (beyond medically assisted reproduction and *in vitro* fertilization), will be able to control his genetic code in the future, adjust his biological programming, fashion his own body. These possibilities are more than foreseeable; they are already part reality. In this sense, Sloterdijk's text is a salutary warning.

It would be unacceptable to turn this call for clear-mindedness into a tacit consent to any *fait accompli* in the biotechnological domain. Some unforeseeable things will certainly occur in the course of investigations and experimentations. Classical humanism was not prepared to face up to that. From now on it will have to acknowledge that ethical concern cannot content itself with incantations, but must constantly bring itself up to date according to new configurations taken by the growth in techno-scientific power.

Another German thinker, Ernst Jünger, has touched on

these new configurations of Power, discerning in them the announcement of a new age, that of the Titanic.[13] More than a philosophical idea, it is an intuition that becomes clear: man is in the process of liberating cosmic energies of which he is less and less the master.

In Greek mythology, the Titans, deities prior to the gods of Olympia, are the first children of the Earth and the Sky: they represent the untamed forces of nature. If Jünger does not envisage the return of the Titans in the literal sense, he finds in them the mythical and striking expression of the demiurgical character of global technology, close to the move that led his friend Heidegger to detect in this globalized technology the era of the gigantic.[14]

Let us actually look at things as they are: what is a human individual before the power of a thermonuclear bomb, before the complex organization of a megalopolis, before the lightning speed of electronic communication? All of these creations, offspring of the human brain, open up fields of forces that crush us as individuals and could well lead to the unforeseeable. Jünger does not think that rationality can control this immense process to the very end: 'Technology is the magical dance that the contemporary world dances'.[15] Will the next Titans still be humans? Will they be anonymous and insensible powers released into the cosmos in the future? Will the man worthy of this name (the one that Jünger names the *Anarque*, the individual who likes to think he is still free) have only 'recourse to forests' (if any of them still survive)?

Jünger, in visionary mode, does not deliver any definitive response, but all his efforts relay the Nietzschean call of the superhuman.

Four

It may be doubtful whether a call of the superhuman is perceptible among present-day humanity, so preoccupied with its 'standard of living', its security and its comfort. If in this respect one must have recourse to Nietzsche, is not the figure of the Last Man essential? 'The earth has become small, and upon it hops the Last Man, who makes everything small [. . .]. "We have invented happiness", say the last Men with a wink.'[1]

It will also be argued, on another level, that it is not the divine but the superhuman that solicits man. Can man find the dimension of Transcendence in himself? We cannot provide an answer to this question of principle within the framework of this limited work, which explores neither the metaphysical dimension nor that of faith. Remaining at the level of a critical reflection on the evolution of the present-day world, we must state – if only to lament it – that we live (as Malraux said) in the first 'atheistic civilization', understood as a technician civilization that, as such, no longer knows anything sacred and does not ascribe a supreme value to itself other than its own efficiency. What a contrast between this absence of a universally shared Transcendence and the constant communication of so-called primitive humanity with the divine powers that succour or threaten it! On the other hand, what is radically missing from our modernity is 'the absolute space of the myth that is identical to the irreducible essence of the gods'.[2] Modern man has irremediably lost his divine models, of which Nietzsche (thinking of the Homeric gods) remarked: 'Man thinks of himself as noble when he bestows upon himself such gods'.[3]

It is in this context of a loss of the sense of the divine and of the death of God (in the Nietzschean sense) that the question of a possible call of the superhuman arises. But which superhuman? It has more than one possible face, especially if one takes seriously the earlier comments on how the techno-scientific world is heading towards the unforeseeable. Distinguishing, anew, between myth and reality does not imply that we make do with demystifying the former to the advantage of the latter (as if it were ever purified of all fantasy). The situation is more complex: mythical repetitions under different forms *signify* something, in the very midst of what we regard, too globally, as our contemporary 'reality'. There is a call to hear and a desire to interpret. And, if it turns out that it is a matter of the metamorphosis or the avatar of the divine, we must put to the test this move towards the extreme and must not be afraid to take it to its limits, tracking down its possible lapses into the inhuman.

> Nietzsche rediscovered the point at which man and God belong to one another, at which the death of the second is synonymous with the disappearance of the first, and at which the promise of the superman signifies first and foremost the imminence of the death of man.[4]

In these dense lines, Foucault has issued an invitation that we must accept: to question Nietzsche on this point 'at which man and God belong to one another'.

Here as elsewhere, Nietzsche's beautiful expressions – however brilliant they are – conceal formidable difficulties. To point to a conflict: must we recognize as direct a connection between the 'promise of the superman' and the 'imminence of the death of man' as Foucault suggests? A promise is not an assured future. Even acknowledging that the death of man is a recognized fact (but to the letter? At what level? In what way?), how does this death predict with certainty an over-coming of the human? Is a regression into the inhuman – alas! – not also conceivable? As it happens Foucault does not answer any of these questions. Even so, one of them can be clarified: the status of the Superman, the nature of the call of the superhuman according to Nietzsche.

THE SUPERHUMAN AS TRANSFIGURATION

'I teach you the Superman. Man is something that should be over-come. What have you done to overcome him?'[5] It is in these terms that Zarathustra addresses the crowd, as soon as he begins his sermon, having only just come down from his retreat in the mountains.

The announcement of the Superman is met at once with the sarcastic remarks of the crowd, who here represent the mediocrity of the Last Men. Nietzsche, then, fosters no illusions about the chances of seeing the superhuman spread across the earth he cherishes ('The Superhuman is the meaning of the earth').[6]

Even if it is not immediately realized and appears beyond the reach of most men, can the superhuman be characterized? How are we to design the route that could take man beyond himself, just as – in the distant past – apes were surpassed by human beings? If it is not possible to go into the detail of the debates between interpreters of Nietzsche here, we can nevertheless point out what is at stake in them: either we see in the Superman the future ruler of the earth, the supreme subject who will realize 'the pure exercise of the will to power', or else we think him outside all typology, freed from technical and political domination, as 'the final and recapitulative individual that is both the oldest and the youngest (or rather the most childlike) of living things',[7] an ultimate figure driving and dominating history or a sumptuous work of art and life extending beyond and defying that same history?[8]

Whether we adopt the first or the second reading, man – according to Nietzsche – should be overcome. He is only a bridge between animality and the divine. He must transfigure his being. And he will be able to undertake this only by focusing his creative will on a goal that, in return, transfigures our reality: the reconciliation with fate, the love of eternity.

Nothing more noble can be contrasted with this Nietzschean call, as viewed from this angle. But is it strictly speaking a *call of the superhuman itself*? No person, no god calls man: the dimension of the superhuman is still a void in which the future gapes open. It is given to the will of man himself (whether it is understood as will to power or abandonment to the innocence of destiny). It is thus man alone who decides to go beyond himself, to draw from his own abyss the best and the worst in order to transgress the hitherto accepted and recognized values. Zarathustra acknowledges, moreover, the disgust he feels for the man who contents himself with persevering in his being.[9] 'To go further', 'to cross the bridge' – these metaphors always signify and announce adventure, exploration, the creative future: 'But he who discovered the country of "Man", also discovered the country of "Human Future". Now you shall be seafarers, brave patient seafarers'.[10]

Transfiguration or disfiguration of the human? We are

obliged to note that there was not only one noble reading of Nietzsche: his radical questioning of the norms of good and evil and his praise for an unconditionally creative will were also exploited in order to cover the worst excesses with an aesthetic alibi. The unavoidable fate of an explosive work that is not unaware that it illustrates the adage: *corruptio optimi pessima* ('the corruption of the best is the worst').

THE SUPERHUMAN AS VERTIGO

The vagueness of the superhuman adds to its fascination. We could almost say that our civilization has no god other than the future, inasmuch as it is constantly progressing technology that seems to promise the impossible. What man finds himself dreaming of (as we saw thanks to some examples drawn from literature and science fiction), is an era where he would have both a titanic power, increased physical and intellectual capacities, and an indefinitely prolonged life in comfort and good health. What could be better? But would this future 'beyond' our unfortunate current limitations really be superior? Would it open up the lives of very different beings?

When we closely examine most utopian projections that use technology as a prop, we are often disappointed to discover only the banal, indeed base, effects of mediocre aspirations. So, to confine ourselves to the desire for immortality, the technologies of plastic surgery, of 'rejuvenation' medicine, indeed of the gruesome preservation of corpses for a hypothetical survival (as in California) appear rather pathetic when faced with the fundamental and constituent reality of the human condition: mortality. Even if one assumes that a day will come when cloning techniques allow us to achieve infinitely superior results, the question will remain: why would we want to survive for such a long time, or for

On the Human Condition

eternity? Do the fabulously wealthy old people who cling to life in sumptuous old people's homes in Florida or California provide an image or even a timorous prefiguration of super-humanity? No, because their survival obsession, pitiful and pathetic, is devoid of any higher motive, any ideal, all enthusiasm. A humanity that has no horizon other than the amassing of quantitative results or the purely technical increase in its physical and mental capacities collapses, loses all energy, is no longer even equal to what, for centuries, was day-to-day human existence, poor and plain, struggling for a dignified tomorrow. Yet, whilst making due allowances for differences, could what we find lamentable about rich ego-istic old people be transposed to a great many aspects of the world of sport, in its profit-making and obsessional version, in which the sole horizon of 'overcoming' is the fanatical gain of some tenths of a second in a race, where young sportsmen and women are ready to undergo dangerous (and illegal) courses of treatment in order to dominate in competi-tions and where, despite this, a champion will be all but worshipped as a superman?

Excessively disillusioned reflections? Perhaps. But we must be conscious of the fact that the growth of capacities and technical performances does not in any way guarantee humanity psychological and moral progress. Even assuming that new technologies improve or prolong human life, what are we to focus on in this progress, if its beneficiaries are increasingly selfish, mean-minded, without imagination or talent? At present everything indicates that we are heading in this direction (or towards this nonsense) of enormous powers for a humanity (without art, religion or philosophy) which has lost every reason to live apart from its own preservation. And everything indicates that this already noticeable difference in a considerable number of individuals will be found increasingly on the global scale: between rich and priveleged humanity (vertiginously devoid of all spiritual life) and the rest: neglected, destitute, but more mobile, unpredictable and close (in their ideas) to their modestly human ancestors.

It must be made quite clear that the prevailing leniency in the 1960s towards utopias was delusory. The utopia of an overcoming of the human is fraught with inhumanity. Stalinism was a utopia (that of an entirely new man), and so too Nazism (the utopia of a racially pure and militarily invincible humanity). Technicalization is another, apparently neutral, utopia, but still more dangerous perhaps in the end, because it does not cease to renew its attractions and disguise what is at stake in them with instant gratifications, the true consequences of which are hidden.

The call of the superhuman is clearly ambiguous: between a nobility reserved for a few and the mirage of technological transformations whose fascinations furnish so many excuses for all moral or intellectual abdications. Should we not return as a matter of urgency to a little more restraint, if the most probable penalty for these 'overcomings' is the regression into new forms of the inhuman?

THE INHUMAN AS ABYSS

The inhuman is not always the penalty for a superhuman excess ('he who wants to play the angel plays the beast'). It can be the product of rapaciousness, of cruelty or quite simply of stupidity. But, whatever its origin, the result can be called degrading, shameful, barbaric.

Here we reach a limit of language and even of what can be thought: how is a man capable of an infinite, virtually absolute, evil? This limit applies to the very epithet 'inhuman' and what it is supposed to denote. Its defect is to let us implicitly believe that this reverse side of the human can and must send us back to its obverse; that there is the possibility of a return from the depths of inhumanity. Such is the hope that is read in the eyes of a future victim seeking a remnant of pity from his or her captor or torturer. Unfortunately, this hope is often disappointed in acts of aggression that are rightly named 'pitiless'. It is this heartless, limitless inhumanity that we have to consider.

If it is true that the inhuman takes us back in principle to the human, dreadful 'experiences' have shown that the inhuman can go as far as distorting the human irredeemably: extermination in its most extreme forms, in the Nazi concentration camps, led to a fatigue of the will in the weakened victims and even the disappearance of the desire for (or the memory of) something like human dignity. Then, as Primo Levi has shown, the human face is not recognizable; man despairs of the human; his body no longer responds; he loses his voice; his eyes grow dull. The torturers thought of themselves as merciless, the victims lost all hope.[11] And we say to ourselves: 'How was that able to happen among humans? Is the return of such horrors really ruled out? How were men able to go as far as these diabolical extremes where the reversal from inhumanity to humanity is no longer even possible?'

What the reading of If This is a Man actually reveals is not only the insane racist cruelty of the Nazis, their mania for sadistic organization, the methodical programming of the degradation of their victims, but also the victims' internalization of their absolute misfortune, the loss of every psychological and moral trait other than immediate survival, the collapse of customs and social instincts short of good and evil:

> The work of bestial degradation, begun by the victorious
> Germans, had been carried to its conclusion by the Germans
> in defeat. It is man who kills, man who creates or suffers
> injustice; it is no longer man who, having lost all restraint,
> shares his bed with a corpse.[12]

The Nazis, who looked upon the Jews, the Slavs, the Hungarian gypsies, etc. as 'sub-human', had no goal other than to make their victims actually regress – if they did not exterminate them – to a subhuman level of brutish survival. But the problem is that the phrases that we have just used are themselves misleading: in this descent into hell, there is no limit other than death.

The inhuman is not a place from which we can easily return. We do not rebound from the inhuman, we sink into it, we get lost in it. It is an abyss. If we take seriously the lessons of these ghastly laboratories – depicted and analysed by Primo Levi without any exaggeration or dramatization, and also without any morbid indulgence – we have to acknowledge that the civilized and moral humanity, the open and welcoming humanity, the humanity in which we would like to recognize ourselves (nay, comfort ourselves) is extremely fragile. As a result, a clear look at 'human nature' must also notice the depths it borders, and how difficult, and even impossible, it becomes to re-establish a human exchange from the moment that an alienating insanity is imposed, as in Primo Levi's depiction of the examination to which he was subjected at Auschwitz:

> Because that look was not one between two men; and if I had
> known how completely to explain the nature of that look,
> which came as if across the glass window of an aquarium
> between two beings who live in different worlds, I would also

have explained the essence of the great insanity of the Third Reich.[13]

One must go so far as wondering whether our impoverished words, starting with the noun 'man', are not incapable of naming what conceals itself in us, both on the side of the superhuman and the inhuman – the two unfathomable extremes of our condition. That is why we must not think about the possibilities of that condition in *opposition* to humanism, but we must do so in a *better* way. It is only in this way that we can succeed in not falling back into the misunderstandings and confusions found in most of the debates on humanism.

Conclusion: What 'Overcoming' Means

Employing truthful language means that we yield neither to edifying pessimism nor the opposite: the optimism that goes with technologization. Humanist discourse does not secure anything; nor does its defeat.

Ought we to add our voices to the alarmist warnings about reproductive cloning or the dangers of eugenics? Vigilance is most certainly vital, because it is not impossible for research (sustained by the profit motive) to produce monstrous things. But, on the other hand, it is unlikely that, in a foreseeable future, man will cross thresholds that amount to escaping his condition. To a very large extent, the overcoming of the human is a myth – a myth favourable to the rapid development of science and technology, a myth furthered by that promotional inflation that we have called 'techno-discourse'. Why is this myth so powerful? Because man is himself an overcoming, but always in an ambiguous sense. What awaits the humanity of tomorrow is not superhumanity, despite the Titanic elements of technological development: these are new forms – unfortunately quite possibly ghastly – of inhumanity. The warnings, when they are stern, look too pessimistic or even cynical. How comforting it is to hear only hymns to progress and glorious tomorrows! It is too easy to soothe one's conscience by becoming indignant about, for example,

Sloterdijk and the *Rules for the Human Park*! But just as the latter has shown, the future of man (who does not have a definitive existence, but 'has to make himself in a permanent quarrel about his non-determined being') is not only mortgaged by biotechnological transformations that are in preparation, it may be still more so by the new forms of violence that are produced by the technician-civilization and spread by the media (all sorts of terrorism, gratuitous violence, mass psychological release, sadomasochism and perversions broadcast by the internet, cinema, television, video, etc.).[1] These prospects are worrying and they tend to alarm us in accentuating the too often recognized powerlessness of the 'classic' means of defence (the family, school, humanist culture, conventional systems of control and sanction) against such outpourings. How are we to face up to things on the two fronts: reacting effectively and seeing what is really happening?

In accordance with this balancing act, similar to what Bergson designated 'double frenzy', we can certainly be tempted – through personal taste, through 'realism' or in the name of progress – to systematically take the opposite view to humanist warnings: 'Science and technology have already worked such wonders: let us leave them to solve the problems posed by their progress!'[2] Some – like the Raëlien sect – make almost a religion of it, pushing scientism and technologism as far as to provoke; others content themselves – quite an achievement – with speaking in praise of scientific and technological advances, framing them with a halo of systematic optimism. The enthusiasm wins over even the specialized scientific set when a new field of research opens up possibilities undreamt of up to that point (this is currently the case with nanotechnologies).

Yet, surely, if one likes to think of oneself as something of a philosopher, or if one wants very simply to retain a few remnants of common sense and clear-mindedness, would it be possible to put prejudices and temperament aside and try honestly to surmount this 'double frenzy', whether humanist or technicist?

That is what we are attempting here. To end up with what results? To obtain a little more clarity in thought and action thanks to an essential distinction between description and prescription, between ontology and deontology, and between the realities to accept and the ethical requirements to reaffirm.

The division between inhuman and superhuman actually corresponds to the two fronts on which man, this chronically unstable being, struggles to stabilize his existence: between inhuman regression and superhuman overcoming, between bestiality and angelism, between malevolence and deification. It should be emphasized that Pascal knew how to lay out the unstable and always surprising territory of the human, that 'in between' that results in 'man infinitely surpass[ing] man', but that can also lead him to collapse into a calamity or a malevolence worse than bestiality.[3]

What ethical response are we to set against this earthquake – this fissure between dangers and call – where everything could become confused as a result of the crisis of our civilization, and the collapse of its traditional values? That response can be as resolute as it is subtle, on condition that it is the result of an understanding of the new division of our humanity, turned enigmatic in its own eyes. What is certain (we have established the truth of it) is that human identity has been subverted and that we can expect astonishing symbioses between man and machines, new forms of the control and manipulation of life and radical moves forward in the global destiny of humanity.[4]

We must defend ourselves against barbaric regressions (this is not easy, because humanitarian action – with its merits and its ambiguities – is a very fragile barrier against the surges and sophistications of evil when evil takes on the extreme and subtle forms that reveal the leading astray and squandering, by man himself, of his own faculties of creation and overcoming).[5] A preventive humanism is essential. How can we not be humanist before the horror of the extermination camps, before the madness of terrorist attacks on innocent people? This is the primary front that we must try to hold, in the urgency of actual practice.

However, limiting ourselves to this necessary ethical response is not sufficient, because it does not take place at the level of the excess that inhabits man to the point where it renders him enigmatic in his own eyes. It means giving up thinking the unthinkable in man, the opening onto this great space, the call of the superhuman when it really seems to raise man above himself. That is the other horizon to be explored – a task allotted to philosophical and poetic thought.

Thus we must know how to open our minds to the possible – provided that it is not regressive. Is ethics outmoded then? Yes and no. Yes, inasmuch as its prescriptions – however legitimate and necessary they are – must not block the horizon of unrestricted and probing thought. No, inasmuch as there are things which could never be justified in the name of the requirements of scientific research (established as an end in itself, most often subject, in fact, to economico-political exigencies of 'competitiveness').[6]

A humanity that stopped wondering about itself would cease to be free. Three hundred years ago, without having need of all our technological marvels to arrive at this intuition, the brilliant Pascal saw right through the irreducible ambiguity of the human condition, its instability and its balance between extremes (destitution, greatness), without sustaining the illusion of definitively warding off this always rekindled, sometimes unbearable, tension between the beast and the angel.

If one takes account of this teaching, topical as never before, thought and action must finally and henceforth be in a position to confront this fluid human complexity. We must know how to establish, as a matter of urgency, a paradoxical 'economy' strategically combining a *cautious humanism*, warning against the inhuman or the subhuman, and an *opening up to possible* super-humans (or everything other than the human 'all too human': disturbing, strange, radically creative) that lie dormant in us. On the one hand, the *defence* of the human against the inhuman, on the other, the *illustration* of what surpasses the human in man.

Without this double strategy, do we not oscillate between moralism and technicalism? Do we not remain shaken about or torn to pieces by the 'double frenzy' exposed by Bergson?

How are we to navigate between the cautious humanism and the daring opening to the possible so as to avoid this 'double frenzy'? How are we to put into practice a strategy that is fruitful and worthy of human greatness (indissociable from its destitution)? Thanks to the understanding of our mortal and fragile division, by accepting our paradoxes and our complexity, by taking on the weight of our entire freedom safeguarded until now.

Why not attempt the impossible: to succeed in crossing ethical vigilance with the call of the superhuman? In this way we will not be mistaken about 'beyond', as we will not have tried to find substitutes for our freedom.

INTRODUCTION   THE OVERCOMING OF OVERCOMING

1 For a more detailed overview of Janicaud's life and work, please see my obituary that appeared in *Research in Phenomenology*, Vol. 33 (2003), pp. 3–5. See also Jean-François Mattéi's useful obituary in *Revue philosophique*, no. 2 (2003), pp. 267–268.

2 *L'homme va-t-il dépasser l'humain?* (Paris: Bayard, 2002).

3 *La métaphysique à la limite* (Paris: Presses Universitaires de France, 1983), p. 23. Translated by Michael Gendre as *Heidegger from Metaphysics to Thought* (Albany: State University of New York Press, 1995). All references are to the French edition and translations are my own.

4 Heidegger, *On Time and Being*, trans. J. Stambaugh (New York: Harper & Row, 1972), p. 24.

5 *La métaphysique à la limite*, op. cit., p. 23.

6 Ibid., p. 23.

7 Ibid., p. 25; the text of Deleuze that Janicaud has in mind is *Différence et répétition* (Paris: Presses Universitaires de France, 1968), p. 90.

8 See 'Author's Preface to the English Edition', *Rationalities, Historicities* (New Jersey: Humanities Press, 1997), p. xiv.

9 *La métaphysique à la limite*, op. cit., p. 31.

10 *La puissance du rationnel* (Paris: Gallimard, 1985). Translated by Peg and Elizabeth Birmingham as *Powers of the Rational* (Bloomington: Indiana University Press, 1994). All references are to the French edition and translations are my own.

11 *Aristote aux champs-élysées* (La versanne: Encre marine, 2003), p. 143.

12 *À nouveau la philosophie* (Paris: Albin Michel, 1991).

13 *La métaphysique à la limite*, op. cit. p. 39. I am not happy with translating 'un

possible' by 'possibility', but see no other possibility. The formulation 'a possible' simply doesn't work in English. As Jean Grondin rightly pointed out in a commemorative conference on Janicaud's work held in Nice in September 2003, 'possibility' connotes something more abstract than the more particular and concrete sense of the possible.

14 *Aristote aux champs-élysées*, op. cit., pp. 159–61.

15 *La puissance du rationnel*, op. cit., p. 321.

16 *Atomised*, trans. F. Wynne (London: Vintage, 2000), p. 377.

17 *L'homme va-t-il dépasser l'humain?*, p. 97.

18 Ibid., p. 91.

19 Ibid., p. 91.

20 Ibid., p. 55.

21 Ibid., p. 100.

22 Ibid., pp. 103–104.

23 Ibid., pp. 100–101.

24 *La puissance du rationnel*, op. cit. p. 342.

25 Ibid., p. 377.

26 Ibid., p. 372.

27 A variant of the argument of my M. Phil thesis found its way into my *Continental Philosophy. A Very Short Introduction* (Oxford: Oxford University Press, 2001), see Chapter 6, 'A Case Study in Misunderstanding: Heidegger and Carnap', pp. 90–110.

28 *Heidegger en France* (Paris: Albin Michel, 2001), p. 443.

PREFACE

1 See Catherine Vincent, *Le Monde*, 10 November 2001, p. 25:

> But where then has the distinctive feature of man gone? The study of chimpanzees, gorillas and orang-utans in their natural milieu, which has constantly intensified for nearly fifty years, radically calls into question what we have long believed to be the specific characteristics of the human species.

> Mais où est donc passé le propre de l'homme? L'étude des chimpanzés, des gorilles et des orangs-outans dans leur milieu naturel, qui ne cesse de s'intensifier depuis près d'un demi-siècle, remet radicalement en cause ce que l'on crut longtemps être les caractères spécifiques de l'espèce humaine.

ONE IS HUMANISM THE LAST RESORT?

1 See Jean-Claude Guillebaud, *Le Monde diplomatique*, August 2001, p. 20. Faced with the threat of a world-wide triple revolution (economical, computing and genetic), Guillebaud raised the key question: 'Is man becoming extinct?' He updated the recourse to humanism, not without arguments, in his *Principe d'humanité* (Paris: Le Seuil, 2001). You will see that we agree with his practical objective (to defend the dignity of man), but that it does not appear to us that the horizon of his thought need be limited to setting humanity up in 'principle', having the will and 'the power to do so' (op. cit., p. 380). This horizon remains too metaphysical and above all excessively anthropocentric.

2 The second meaning specified by *Le Robert, Dictionnaire alphabétique et analogique de la langue française*: 'toute théorie ou doctrine qui prend pour fin la personne humaine et son épanouissement'.

3 Martin Heidegger, 'Letter on Humanism', trans. Frank A. Capuzzi in William McNeill (ed.), *Pathmarks* (Cambridge: Cambridge University Press, 1998), p. 262.

4 Jean-Paul Sartre, *Existentialism and Humanism* (London: Methuen, 1948), p. 55.

5 Ibid.

6 Ibid., pp. 55–56.

7 Ibid., p. 55.

8 For a more comprehensive exposition of these positions of overstatement, see my recent contribution: 'L'humanisme: des malentendus aux enjeux', *Revue philosophique de Louvain*, May 2001, pp. 183–200.

9 Claude Lévi-Strauss, *The Savage Mind* (London: Weidenfeld and Nicolson, 1966), p. 247.

10 See Mikel Dufrenne, *Pour l'homme* (Paris: Le Seuil, 1968).

11 On this point, see Louis Althusser, *For Marx*, trans. Ben Brewster (London: The Penguin Press, 1969), pp. 221–241.

12 In J. J. Rousseau, *Essai sur l'origine des langues*, VIII, cited in Lévi-Strauss, *The Savage Mind*, p. 247.

13 Claude Lévi-Strauss, 'The three humanisms' in *Structural Anthropology* Vol II, trans. by Monique Layton (London: Penguin Books, 1977), pp. 271–274.

14 Ibid., p. 272.

15 Indeed, Lévi-Strauss presents the humanism that he extols as 'demo-

cratic', inasmuch as he seeks his inspiration 'in the most humble and despised societies', ibid., p. 274.

16 Ibid.

17 See Michel Foucault, *The Order of Things* (London: Tavistock Publications, 1970), pp. 307–312.

18 Ibid., pp. 340–343.

19 Peter Sloterdijk, *Règles pour le parc humain*, trans. O. Mannoni (Paris: Mille et une nuits, 1999).

20 The sale of ova is already taking place in California. See Dominique Dhombres, 'Ovules sur catalogue', *Le Monde*, 16 January 2002, p. 32.

TWO THE DANGER OF MONSTERS

1 See Gabriel Marcel, *Man Against Mass Society* (Chicago, IL: Henry Regnery Company, 1952), pp. 153–162. Reviewing the new techniques of degradation and the ravages of fanaticism, Marcel deserves credit for exposing 'the spirit of abstraction, as a factor making for war'.

2 This definition is found in the dictionary, *Le Robert*: 'un organisme de configuration insolite'.

3 See Georges Canguilhem, *La connaissance de la vie* (Paris: Vrin, 1975), p. 183. Canguilhem notes: 'Life is short of monsters'.

4 Mary Shelly, *Frankenstein or the Modern Prometheus* (London and New York: Penguin Books, 1994), p. 140.

5 Ibid., p. 141.

6 Jean-Michel Truong, *Totalement inhumaine* (Paris: Les Empêcheurs de penser en rond, 2001).

7 This is a hypothesis already formulated by Bill Joy, cited by Edgar Morin, *L'identité humaine* (Paris: Le Seuil, 2001), p. 232.

8 See Jean-Michel Truong, op. cit., pp. 49–50.

9 See Edgar Morin, *Le paradigme perdu: la nature humaine* (Paris: Points-Seuil, 1973), p. 211: 'It is not the notion of man that dies today, but an insular notion of man'. 'Ce qui meurt aujourd'hui, ce n'est pas la notion d'homme, mais une notion insulaire de l'homme.' We no longer have to content ourselves with thinking about human beings starting from [these beings] themselves and their supposedly permanent characteristics, but [we have to] resituate them in the evolution of the living, with respect to their environment as well as in their ethnic and sociocultural differences.

10 Frederik Pohl, *Man Plus* (London: Millennium, 2000).

11 Ibid., p. 189.

12 Ibid., p. 213.

13 See Michel Alberganti, 'L'être bionique, mi-vivant, mi-machine, sort des limbes', *Le Monde*, 20 August 2001, p. 13.

THREE   FROM FORESEEABLE RISKS TO
THE UNFORESEEABLE

1 I suggested the term, 'techno-discourse' in *La puissance du rationnel* (Paris: Gallimard, 1985), p. 100ff.

2 Total valid at the end of 1999. See Axel Kahn, *Et l'homme dans tout ça?* (Paris: Nil Éditions, 2000), p. 225.

3 Ibid., p. 237.

4 Henri Atlan, 'Clonage thérapeutique: gardons-nous des fantasmes', *Le Monde*, 18 January 2002, pp. 1, 14 and 15. See also, *La science est-elle inhumaine?* (Paris: Bayard, 2002), pp. 77–86.

5 Jean-Yves Nau, 'Les promesses des cellules souches concurrencent celles du génome', *Le Monde*, 29 August 2001, p. 17.

6 Marc Roche, 'La Grande-Bretagne veut devenir l'Eldorado des recherches en biotechnologies', *Le Monde*, 17 August 2001, p. 2.

7 A suspicion formulated in the following terms by President Chirac in February 2001:

> I am not in favour of authorising therapeutic cloning. It leads to the creation of embryos for research purposes and for the purposes of cell production and, in spite of the prohibition, renders reproductive cloning physically possible and could easily lead to trafficking in oocytes.

> Je ne suis pas favorable à l'autorisation du clonage thérapeutique. Il conduit à créer des embryons à des fins de recherche et de production de cellules et, malgré l'interdit, rend matériellement possible le clonage reproductif et risque de conduire à des trafics d'ovocytes.

See *Le Monde*, 28 November 2001, p. 26.

8 Jean-Yves Nau, 'Le docteur Antinori lance un programme de clonage humain reproductif', *Le Monde*, 7 August 2001. See also the editorial of 9 August 2001: 'Merci, docteur Antinori'.

9 See the two titles on p. 5 of *Le Monde* dated 9 August 2001: 'Le premier

projet de clonage reproductif humain unanimement condamné'; 'Interdisant ces pratiques, le droit international reste impuissant à les prévenir'.

10 Axel Kahn, *Et l'homme dans tout ça?*, p. 230.

11 See Clifford D. Simak, *Time and Again* (London: Heinemann, 1956); Isaac Asimov, *I, Robot* (London: Dobson, 1967).

12 This lecture, delivered at the end of the Summer of 1999, was first published in French in a supplement to *Le Monde des débats* in October 1999, then in Les Éditions Mille et une nuits in 2000 in Olivier Mannoni's translation.

13 See Ernst Jünger, Antoni Gnoli and Franco Volpi, *Les prochains Titans*, trans. M. Bouzaher (Paris: Grasset, 1998).

14 See Martin Heidegger, *Off the Beaten Track*, trans. Julian Young and Kenneth Haynes (Cambridge: Cambridge University Press, 2002), pp. 71–72.

15 Ernst Jünger, *Les prochains Titans*, p. 19.

FOUR BETWEEN THE SUPERHUMAN AND THE INHUMAN

1 Friedrich Nietzsche, *Thus Spoke Zarathustra*, trans. R. J. Hollingdale (Middlesex and New York: Penguin Books, 1961) [trans. modified].

2 See Pierre Klossowski, *Le bain de Diane* (Paris: Gallimard, 1956), pp. 47 and 62.

3 Friedrich Nietzsche, *Human, All Too Human*, trans. R. J. Hollingdale (Cambridge: Cambridge University Press, 1986, 1996), p. 66.

4 Michel Foucault, *The Order of Things* (London: Tavistock Publications, 1970), p. 342.

5 Friedrich Nietzsche, *Thus Spoke Zarathustra*, p. 41.

6 Ibid., p. 42.

7 This is the thesis that Martin Heidegger defended. See Martin Heidegger, *Nietzsche, Vol II: The Eternal Return of the Same*, trans. David Farrell Krell (San Francisco: HarperCollins, 1991), pp. 215–216.

8 It is in these terms that Jean-François Marquet expresses an interpretation that is truer to the texts and is now winning acceptance among most Nietzscheans. See 'L'individu chez Nietzsche: décadence et récapitulation', *Bulletin de la Société française de philosophie*, July–September 2001, p. 16.

9 Friedrich Nietzsche, *Thus Spoke Zarathustra*, pp. 232–238 ('Der grosse Überdruss am Menschen').

10 Ibid., p. 230.

11 Primo Levi, *If This is a Man*, trans. Stuart Woolf (London: Abacus, 1987).

12 Ibid., p. 177.

13 Ibid., pp. 111–112.

CONCLUSION

1 Peter Sloterdijk, *Règles pour le parc humain*, trans. O. Mannoni (Paris: Mille et une nuits, 1999), p. 58.

2 Henri Bergson, *The Two Sources of Morality and Religion*, trans. by R. Ashley Audra and Cloudesley Brereton with the assistance of W. Horsfall (Carter, Indiana: University of Notre Dame Press, 1977), pp. 293–298.

3 Pascal, *Pensées* VII, no. 434 (Paris: Éd. Brunschvicg), p. 531.

4 Cf. Edgar Morin, *L'identité humaine* (Paris: Le Seuil, 2001), and in particular Chapter 5 of the 3rd part, 'L'identité future', pp. 229–244.

5 See Françoise Lazard, 'La malle-cabine de l'humanitaire', *Le Monde*, 23 February 2001, p. 27.

6 The competition between laboratories exercising tremendous pressure on political convictions and power could well bypass every ethical moratorium. See 'Le biopouvoir à l'assaut des lois de bioéthique', *Le Monde*, 18 January 2002, p. 14 (an article by a spirited group that includes J. Testart, J.-J. Salomon and M. Tibon-Cornillot).

Bibliography

Althusser, Louis, *For Marx*, trans. Ben Brewster (London: The Penguin Press, 1969).

Asimov, Isaac, *I, Robot* (London: Dobson, 1967).

Atlan, Henri, *La science est-elle inhumaine?* (Paris: Bayard, 2002).

Bergson, Henri, *The Two Sources of Morality and Religion*, trans. R. Ashley Audra and Cloudesley Brereton with the assistance of W. Horsfall (Carter, Indiana: University of Notre Dame Press, 1977).

Canguilhem, Georges, *La connaissance de la vie* (Paris: Vrin, 1975).

Critchley, Simon, *Continental Philosophy. A Very Short Introduction* (Oxford: Oxford University Press, 2001).

Deleuze, Gilles, *Différence et répétition* (Paris: Presses Universitaires de France, 1968).

Dufrenne, Mikel, *Pour l'homme* (Paris: Le Seuil, 1968).

Ferry, Luc and Vincent, Jean-Didier, *Qu'est-ce que l'homme?* (Paris: Odile Jacob, 2000).

Foucault, Michel, *The Order of Things* (London: Routledge, 2001).

Gontier, Thierry, 'Une catégorie historio-graphique oblitérée, l'humanisme', in Y.-C. Zarka (ed.), *Comment écrire l'histoire de la philosophie?* (Paris: PUF-Quadrige, 2001), pp. 267–281.

Guillebaud, Jean-Claude, *Le principe d'humanité* (Paris: Le Seuil, 2001).

Heidegger, Martin, *Off the Beaten Track*, trans. Julian Young and Kenneth Haynes (Cambridge: Cambridge University Press, 2002).

—— 'Letter on humanism', trans. Frank A. Capuzzi in William McNeill (ed.), *Pathmarks* (Cambridge: Cambridge University Press, 1998).

—— *Nietzsche, Vol II: The Eternal Return of the Same*, trans. David Farrell Krell (San Francisco: HarperCollins, 1991).

—— On Time and Being, trans. J. Stambaugh (New York: Harper & Row, 1972).

Houellebecq, Michel, Atomised, trans. F. Wynne (London: Vintage, 2000).

Janicaud, Dominique, Aristote aux champs-élysées (La versanne: Encre marine, 2003).

—— 'L'humanisme: des malentendus aux enjeux', Revue philosophique de Louvain, May 2001, pp. 183–199.

—— Heidegger en France (Paris: Albin Michel, 2001).

—— 'Author's Preface', Rationalities, Historicities (New Jersey: Humanities Press, 1997).

—— La puissance du rationnel (Paris: Gallimard, 1985), trans. Peg and Elizabeth Birmingham as Powers of the Rational (Bloomington: Indiana University Press, 1994).

—— À nouveau la philosophie (Paris: Albin Michel, 1991).

Janicaud, Dominique and Mattéi, Jean-François, La métaphysique à la limite (Paris: Presses Universitaires de France, 1983), trans. Michael Gendre as Heidegger from Metaphysics to Thought (Albany: State University of New York Press, 1995).

Jünger, Ernst, Antoni Gnoli and Franco Volpi, Les prochains Titans, trans. M. Bouzaher (Paris: Grasset, 1998).

Kahn, Axel, Et l'homme dans tout ça? (Paris: Nil Éditions, 2000).

Klossowski, Pierre, Le bain de Diane (Paris: Gallimard, 1956).

Koninck, Thomas de, De la dignité humaine (Paris: PUF, 1995).

Levi, Primo, If This is a Man, trans. Stuart Woolf (London: Abacus, 1987).

Lévi-Strauss, Claude, 'The three humanisms' in Structural Anthropology Vol II, trans. by Monique Layton (London: Penguin Books, 1977).

—— The Savage Mind (London: Weidenfeld and Nicolson, 1966).

Magnard, Pierre, Questions à l'humanisme (Paris: PUF, 2000).

Marcel, Gabriel, Man Against Mass Society (Chicago, IL: Henry Regnery Company, 1952).

Morin, Edgar, L'identité humaine (Paris: Le Seuil, 2001).

—— Le paradigme perdu: la nature humaine (Paris: Points-Seuil, 1973).

Nietzsche, Friedrich, Human, All Too Human, trans. R. J. Hollingdale (Cambridge: Cambridge University Press, 1986).

—— Thus Spoke Zarathustra, trans. R. J. Hollingdale (Middlesex and New York: Penguin Books, 1961).

Pascal, Blaise, Pensées VII, no. 434 (Paris: Éd. Brunschvicg, 1904–1914).

Pohl, Frederik, Man Plus (London: Millennium, 2000).

Sartre, Jean-Paul, *Existentialism and Humanism* (London: Methuen, 1948).

Shelly, Mary, *Frankenstein or the Modern Prometheus* (London and New York: Penguin Books, 1994).

Simak, Clifford D., *Time and Again* (London: Heinemann, 1956).

Sloterdijk, Peter, *Règles pour le parc humain*, trans. O. Mannoni (Paris: Mille et une nuits, 1999).

Truong, Jean-Michel, *Totalement inhumaine* (Paris: Les Empêcheurs de penser en rond, 2001).

Index

Adam and Eve 24
Adorno, Theodor W. 42
Althusser, Louis 13, 16
anthropocentrism, critique of 10
'anthropological sleep' 16
anthropology 5, 13, 15
anti-humanism 13, 15
Antinori, Dr Severino 38
Aquinas, St Thomas 25
Aristotle 6
artificial intelligence 27
Atlan, Henri 37
Attila 19
Augustine St 24
Auschwitz 52

Barthes, Roland 12
Beaufret, Jean 7
being 10, 11, 15, 47
Bergson, Henri 28, 34, 55, 58
Bible 2
bioengineering 1, 32, 42
bioethics 2, 17
biopolitics 41
biotechnologies 17, 32, 38, 41

call of the superhuman 4, 43, 44,
    45–7, 50, 57

Cicero, Marcus Tullius 6
cloning 4, 36–8, 39, 48, 54
consciousness 29–30
Critique of Dialectical Reason 12
cyborg, myth of 30–4

Dasein 9, 15
'death of God' 15, 45
'death of man' 15, 16, 45–6
Delphic oracle 6
Descartes, René 10, 16
Diderot, Denis 14
Dolly, the sheep 36
Dufrenne, Mikel 13
Dumézil, Georges 11
Durkheim, Émile 12

elixir of life 24
emancipatory practice 15
enfant à la carte 37
Enlightenment 10, 16
ethics 3, 17, 18, 36–9, 42, 56–8
ethnology 13–15
evil 20–1, 24, 34, 51, 57
Existentialism is a Humanism 7

Foucault, Michel 13, 15–16, 40,
    45–6

Frankenstein 23–7
Frankfurt School 42
freedom 9, 34

Galilei, Galileo 25
Garden of Eden 30
genetic engineering 17
globalisation 1, 28, 29
God 6, 9, 10, 45
gods 20, 45, 47, 48
Goethe, Johann Wolfgang von 24

Habermas, Jürgen 42
Heidegger, Martin 7–11, 15, 42, 43
Hitler, Adolf 16, 19
human: achievements 8, 11;
    capacities 4, 33; condition 34,
    48, 58; dignity 51; existence 49;
    identity 1, 56; nature 2, 5–6, 8,
    52; possibilities 10, 17, 21;
    species 1, 22
humanism 3, 5–18, 41, 42, 53,
    57–8; and critical reflections on
    3, 5, 8–17
'human park' 18, 41
human sciences 6, 13, 14, 16
Hydra and Cerberus 19

ideology 3, 4, 12, 14, 17
If This is a Man 51
inhumanity 4, 10, 19, 20, 21–2,
    25, 26, 29, 30, 34, 38–9, 45, 46,
    50–3, 54, 56, 58
internet 2, 27, 28, 29
in vitro fertilization 17, 42

Jünger, Ernst 40, 42–3

Kahn, Axel 37
Kant, Immanuel 6,
Kantian moment 16

language 7, 11, 50, 54
last man 44, 46
Leonardo da Vinci 24–5
Letter on Humanism 7, 8
Levi, Primo 51, 52
Lévi-Strauss, Claude 11–15
litterae humaniores 6
Lucretius 6

Malraux, André 44
man 2–17, 19–22, 26, 29–30, 33,
    39–43, 45, 47–8, 50–1, 53,
    54–7
Man Against Mass Society 19
Man Plus 34
Marcel, Gabriel 19
Marx, Karl 9, 13
meaning 6, 7, 11
metaphysical fixation 10
monsters 4, 5, 17, 18, 20–7
morality 16, 22, 25, 38, 50
myth 4, 23, 33, 36, 45, 54
mythologies 6, 19

Nazi concentration camps 51
Nazis 7, 51, 52
Nazism 50
Nietzsche, Friedrich Wilhelm 16,
    28, 33, 42, 44–8
Nietzscheanism 41

ontological difference 10
Order of Things 15

original sin 24
overcoming 2, 17, 27, 33, 39, 41, 46, 49, 50, 54–8
ozone layer 40

Pascal, Blaise 22, 24, 56, 57
plastic surgery 48
poetic thought 57
Pohl, Frederik 30
Pol Pot 16
positivism 8
post-humanity 28
Proudhon, Pierre Joseph 6

Renaissance 6, 10
'return of masks' 16
risk assessment 4, 35–43, 54
robots 27
Rousseau, Jean Jacques 14

Sartre, Jean-Paul 7–11, 12, 13, 15
Saussure, Ferdinand de 11
Savage Mind 12
science 1–2, 14–15, 20, 23–6, 54, 55
science fiction 4, 17, 26, 27, 32, 39, 48
Shelley, Mary 23–5
Sloterdijk, Peter 17, 41–2, 55
Socrates 6

sport 49
Stalin, Joseph 16
Stalinism 50
stem cell culture 37
structuralism 11–13, 17
subjectivity 9, 10, 30
suicide 12, 30
superhuman 4, 26, 32, 33, 44–50, 53, 56
superman 28, 33, 46–9
surrogate mothers 17

'techno-discourse' 35, 54
technology 1–2, 15, 17, 20, 26, 32–4, 41, 43, 48–9, 54–5
Teilhard de Chardin, Pierre 29
terrorism 38, 40, 55
titans 43
transcendence 44
Tristes Tropiques 13
Truong, Jean-Michel 27, 29
'truth of Being' 9

utopias 48, 50
'will to power' 47

Valéry, Paul Ambroise 40

Zarathustra 46–7

THINKING IN ACTION – order more now

TITLE	AUTHOR	ISBN	BIND	PRICES UK	US	CANADA
On Education	Harry Brighouse	0415327903	PB	£8.99	$14.95	$19.95
On the Human Condition	Dominique Janicaud	0415327962	PB	£8.99	$14.95	$19.95
On the Public	Alastair Hannay	0415327938	PB	£8.99	$14.95	$19.95
On the Political	Chantal Mouffe	0415305217	PB	£8.99	$14.95	$19.95
On Belief	Slavoj Zizek	0415255325	PB	£8.99	$14.95	$19.95
On Cosmopolitanism and Forgiveness	Jacques Derrida	0415227127	PB	£8.99	$14.95	$19.95
On Film	Stephen Mulhall	0415247969	PB	£8.99	$14.95	$19.95
On Being Authentic	Charles Guignon	0415261236	PB	£8.99	$14.95	$19.95
On Humour	Simon Critchley	0415251214	PB	£8.99	$14.95	$19.95
On Immigration and Refugees	Sir Michael Dummett	0415227089	PB	£8.99	$14.95	$19.95
On Anxiety	Reneta Salecl	0415312760	PB	£8.99	$14.95	$19.95
On Literature	Hillis Miller	0415261252	PB	£8.99	$14.95	$19.95
On Religion	John D Caputo	041523333X	PB	£8.99	$14.95	$19.95
On Humanism	Richard Norman	0415305233	PB	£8.99	$14.95	$19.95
On Science	Brian Ridley	0415249805	PB	£8.99	$14.95	$19.95
On Stories	Richard Kearney	0415247985	PB	£8.99	$14.95	$19.95
On Personality	Peter Goldie	0415305144	PB	£8.99	$14.95	$19.95
On the Internet	Hubert Dreyfus	0415228077	PB	£8.99	$14.95	$19.95
On Evil	Adam Morton	0415305195	PB	£8.99	$14.95	$19.95
On the Meaning of Life	John Cottingham	0415248000	PB	£8.99	$14.95	$19.95
On Cloning	John Harris	0415317002	PB	£8.99	$14.95	$19.95

...Big ideas to fit in your pocket